walkermaths 3.2

LINEAR PROGRAMMING

NCEA Level 3 Internal

Charlotte Walker and Victoria Walker

NELSON
A Cengage Company

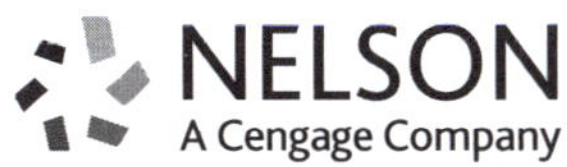

Australia • Brazil • Japan • Korea • Mexico • Singapore • Spain • United Kingdom • United States

Walker Maths 3.2 Linear Programming
1st Edition
Charlotte Walker
Victoria Walker

Editor: Eva Chan
Designer: Cheryl Rowe, Macarn Design
Production controller: Siew Han Ong

Any URLs contained in this publication were checked for currency during the production process. Note, however, that the publisher cannot vouch for the ongoing currency of URLs.

Acknowledgements
Cover photo courtesy of Shutterstock.

For product information and technology assistance,
in Australia call **1300 790 853**;
in New Zealand call **0800 449 725**

For permission to use material from this text or product, please email **aust.permissions@cengage.com**

National Library of New Zealand Cataloguing-in-Publication Data
A catalogue record for this book is available from the National Library of New Zealand.

978 0 17 038939 6

Cengage Learning Australia
Level 7, 80 Dorcas Street
South Melbourne, Victoria Australia 3205

Cengage Learning New Zealand
Unit 4B Rosedale Office Park
331 Rosedale Road, Albany, North Shore 0632, NZ

For learning solutions, visit **cengage.co.nz**

Printed in China by 1010 Printing International Limited.
11 24

CONTENTS

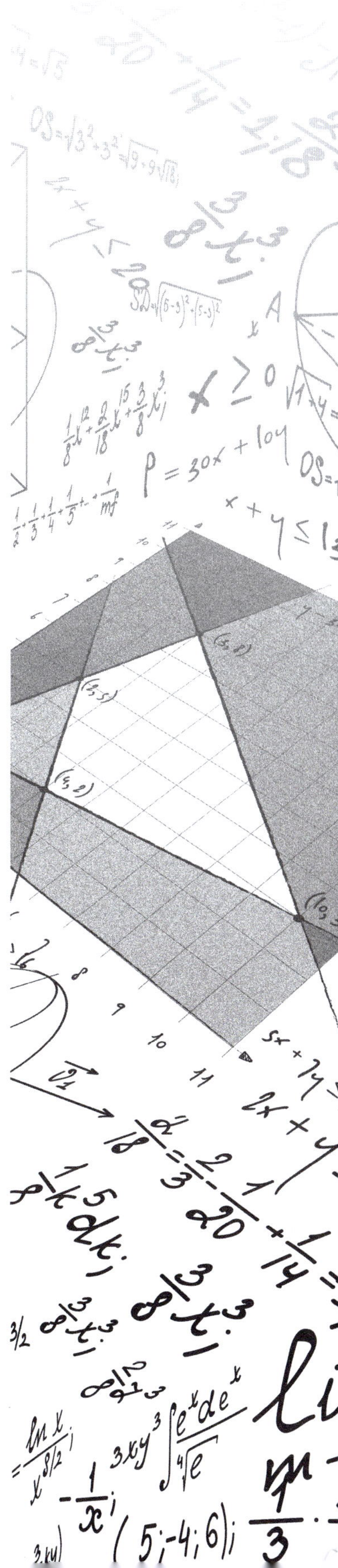

ISBN: 9780170389396

Glossary

Make your own glossary of key terms:

Term	Definition	Picture/Example
Feasible region		
Vertex		
Vertices		
Constraints		
Objective function		
Inequation		
Integral		
Polygon		

What is linear programming?

Linear programming is a process for finding a maximum or minimum value for an objective function that is subject to several constraints. It involves:

- writing and graphing equations for the constraints in order to find a feasible region that satisfies all equations at once
- finding the coordinates of each corner of the feasible region
- evaluating an objective function at each corner of the feasible region
- deciding which corner of the feasible region represents the maximum or minimum value for the objective function.

ISBN: 9780170389396

Drawing straight lines

1 Plotting points using the equation

Example:
Plot the line given by the equation $y = -2x + 5$.
Step 1: Create a table for values of x and y.

Note that this is easiest if your equation is in the $y = mx + c$ form.

x	$-2x + 5$	y	Point
0	-2(0) + 5	5	(0, 5)
1	-2(1) + 5	3	(1, 3)
2	-2(2) + 5	1	(2, 1)
3	-2(3) + 5	-1	(3, -1)
4	-2(4) + 5	-3	(4, -3)

Step 2: Plot the points on a graph.

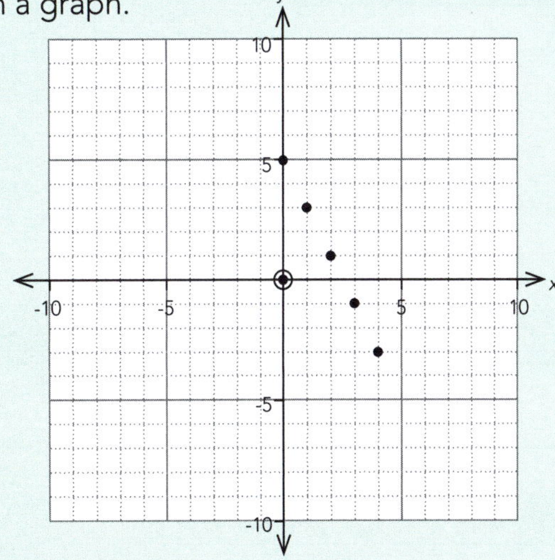

Step 3: Join the points with a **ruled** line.

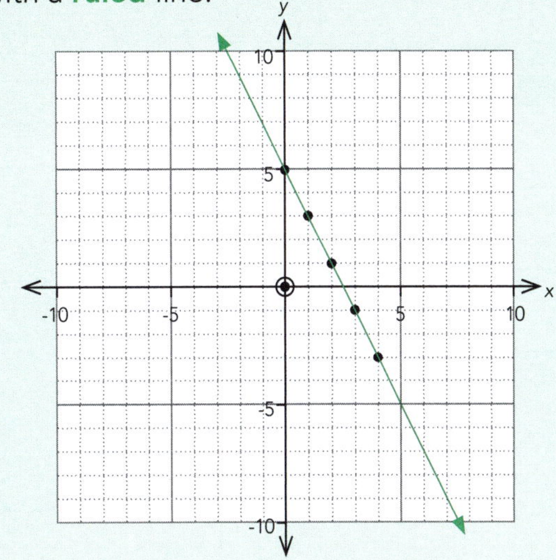

ISBN: 9780170389396

Complete the tables and draw the graph for each of the following.

1 $y = 3x - 2$

x	3x – 2	y	Point
0	3(0) – 2	-2	(0, -2)
1	3(1) – 2		
2	3(2) – 2		
3	3(3) – 2		
4	3(4) – 2		

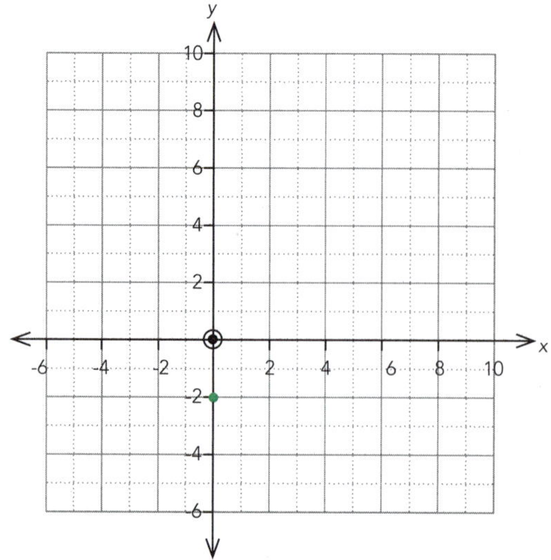

2 $y = \frac{1}{2}x + 3$

x	$\frac{1}{2}x + 3$	y	Point
0	$\frac{1}{2}(0) + 3$	3	(0, 3)
2			
4			
6			
8			

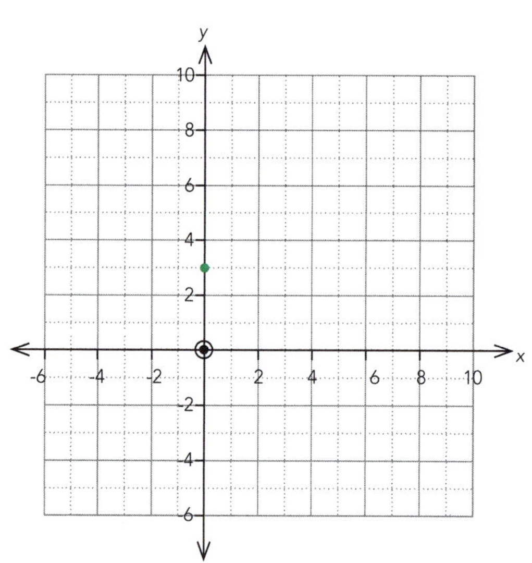

Make substitution easier by using even numbers.

3 $y = -4x + 5$

x	-4x + 5	y	Point
0			
1			
2			
3			
4			

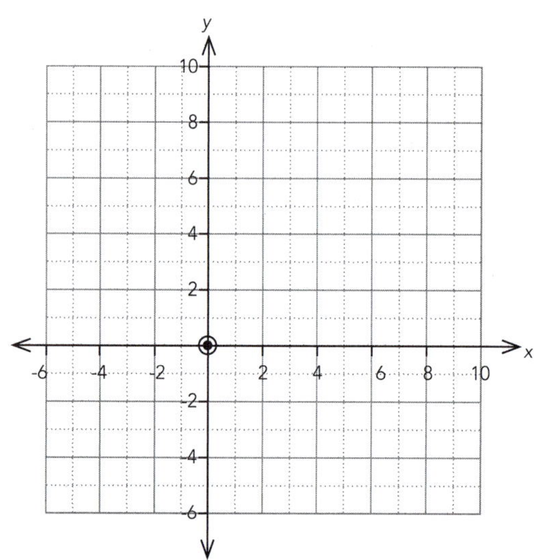

 ISBN: 9780170389396

4 $y = -\frac{1}{4}x$

x	$-\frac{1}{4}x$	**y**	**Point**

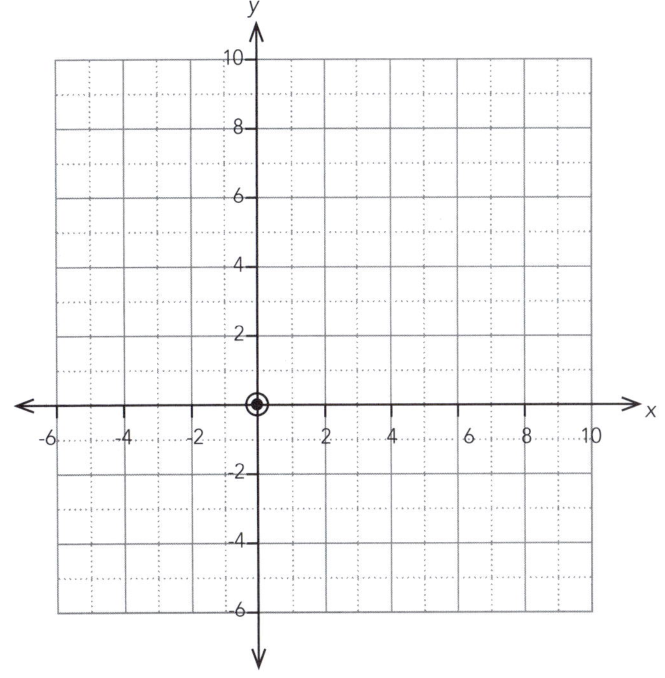

5 $y = -5x - 2$

x		**y**	**Point**

6 $y = \frac{1}{2}x + 120$

x			
0			
100			

You will need to use bigger values for x.

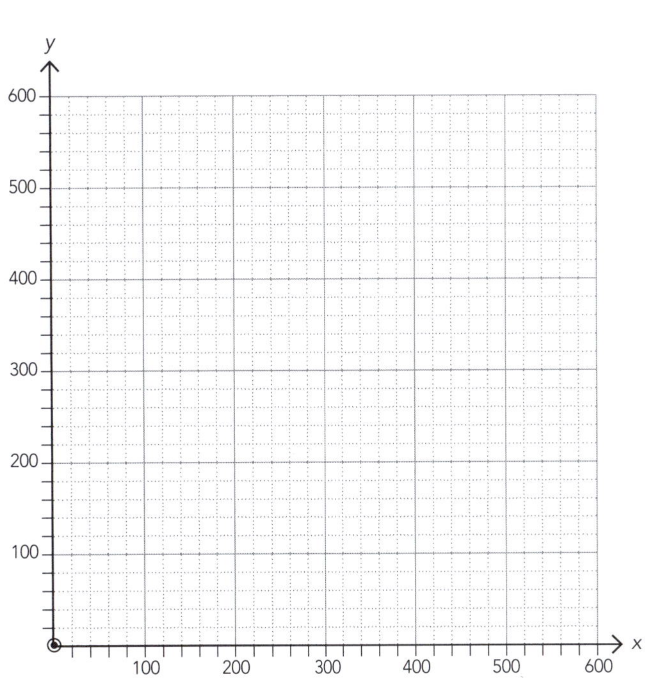

7 $y = 480 - 2x$

			Point

2 Finding points using the table function on your calculator

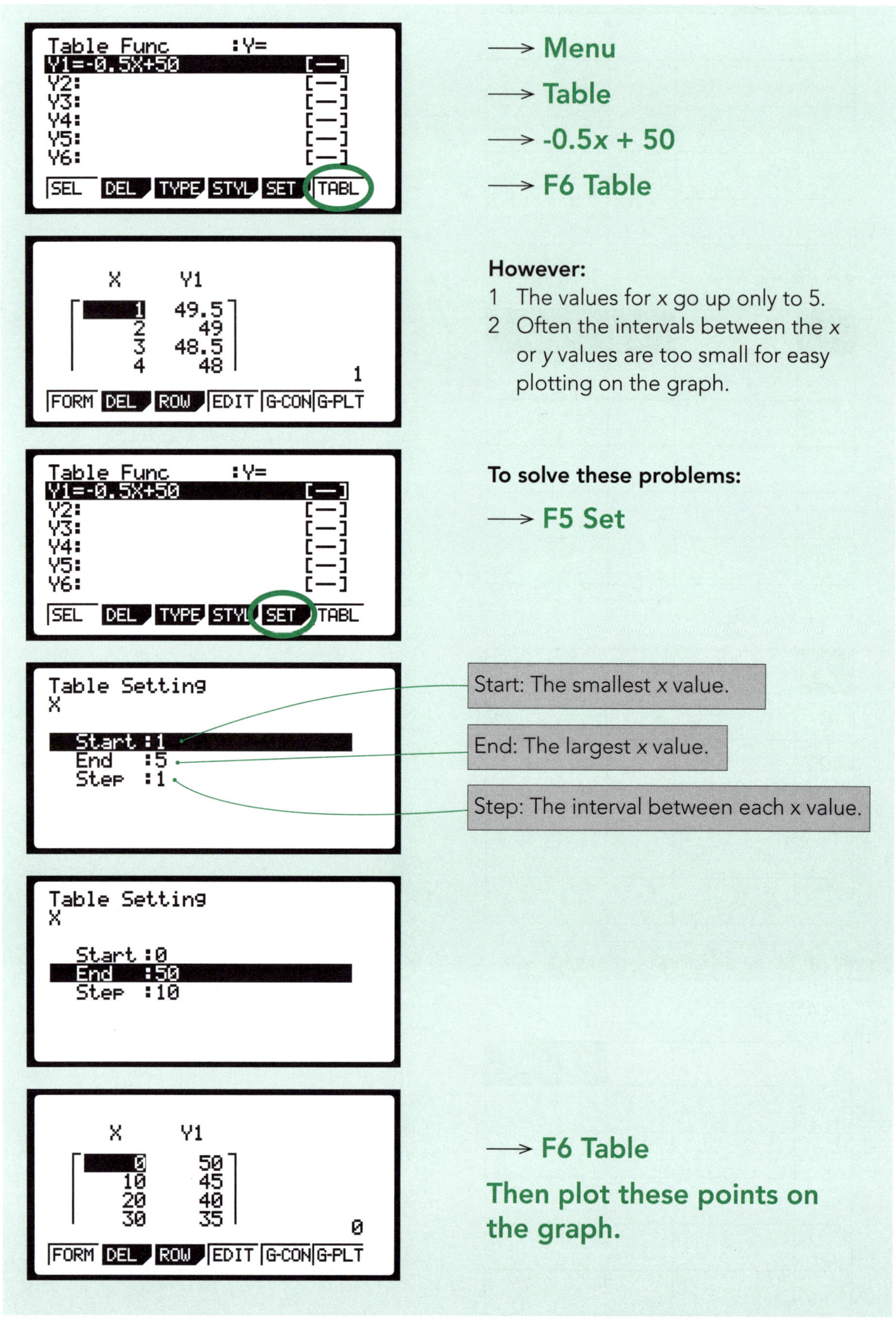

→ **Menu**

→ **Table**

→ **-0.5*x* + 50**

→ **F6 Table**

However:

1 The values for *x* go up only to 5.
2 Often the intervals between the *x* or *y* values are too small for easy plotting on the graph.

To solve these problems:

→ **F5 Set**

→ **F6 Table**

Then plot these points on the graph.

ISBN: 9780170389396

Where necessary, convert the following equations to $y =$ ……, and use your calculator to help you draw the graph.

1 $y = x + 5$

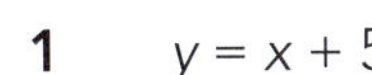

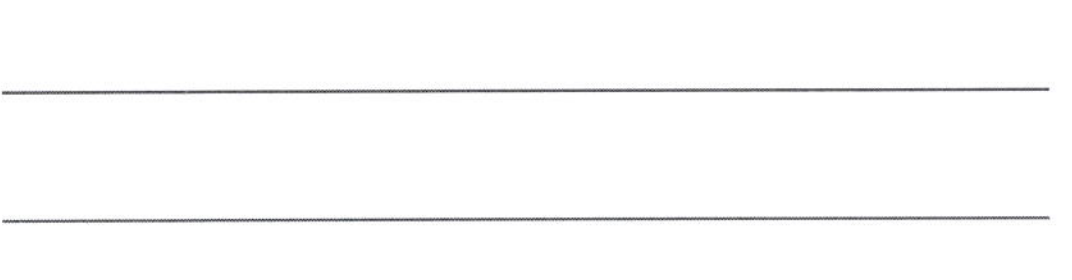

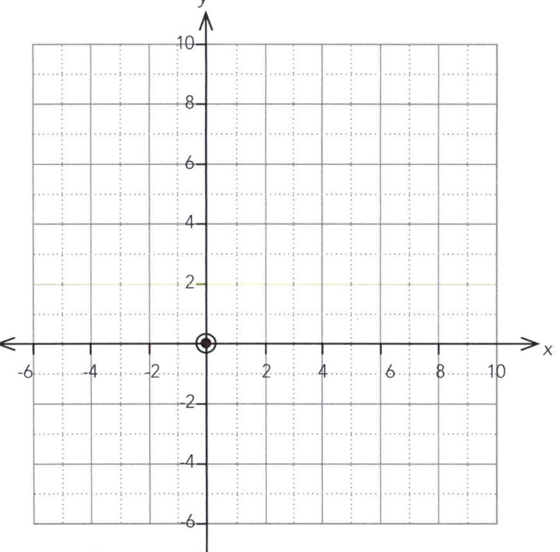

2 $y + 3x = 6$

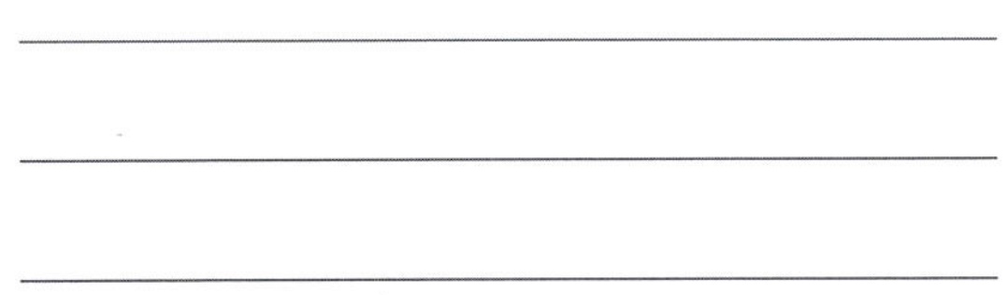

3 $y = 36 - 6x$

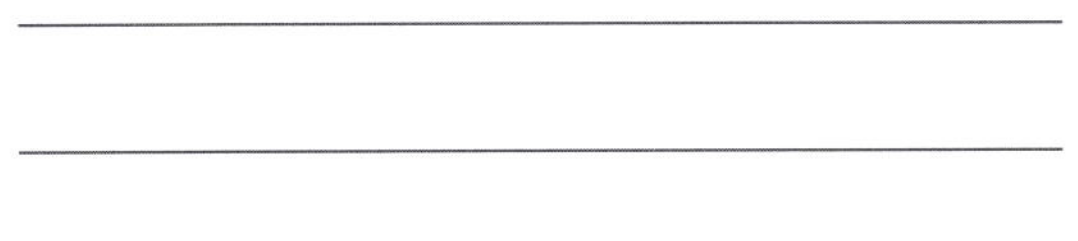

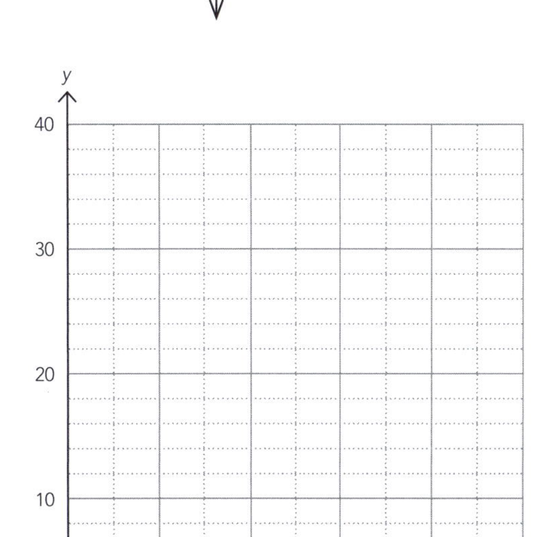

4 $y - x = 6$

5 $y + 5x = 85$

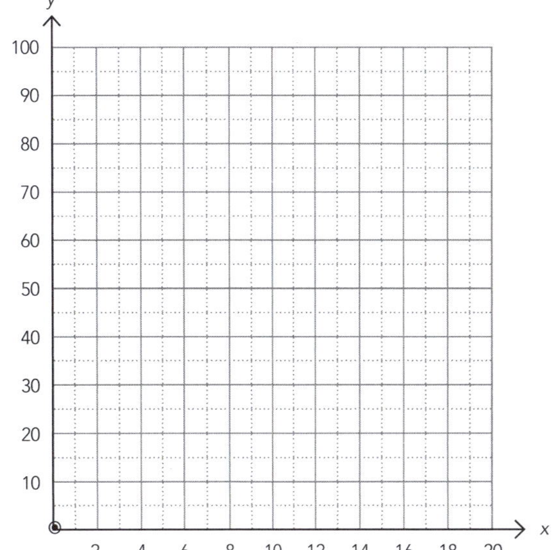

6 $4x - y + 10 = 0$

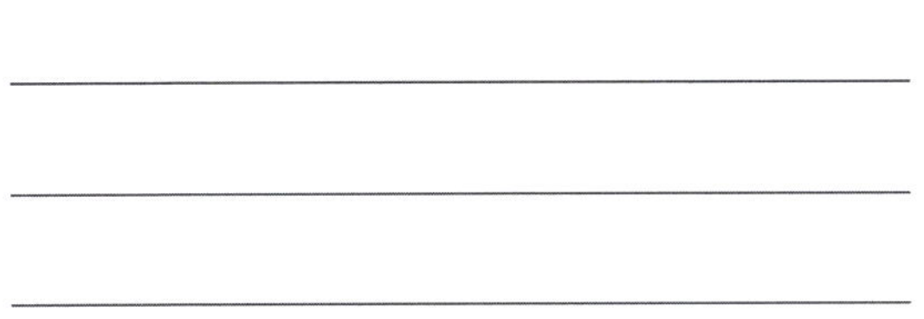

ISBN: 9780170389396

7 $5x + 7y = 2800$

8 $y = 160 + 0.5x$

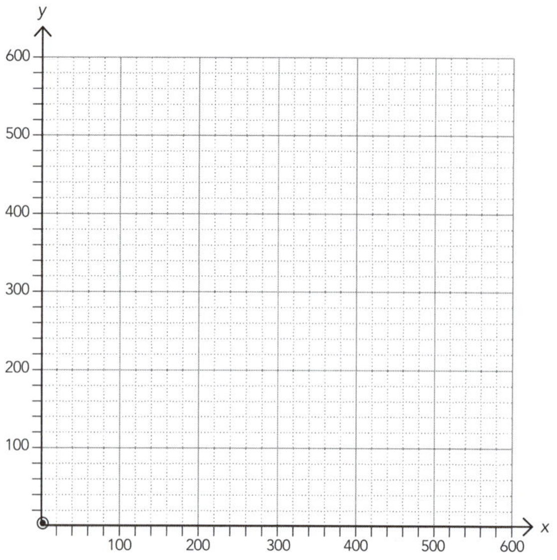

9 $5y - x = 100$

10 $x + 3y - 540 = 0$

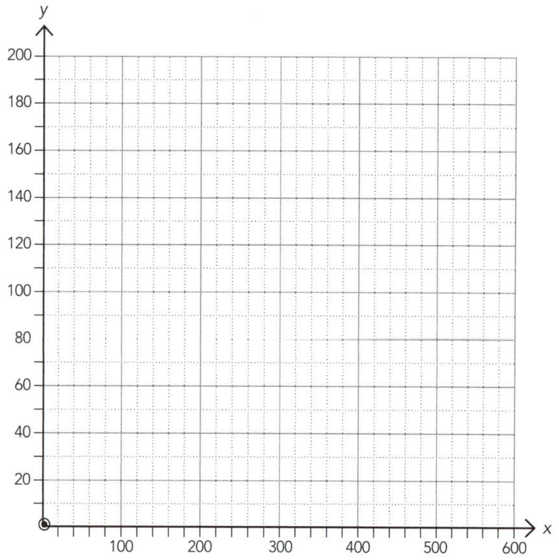

11 $a = 3b$

12 $1200 - 5a - 6b = 0$

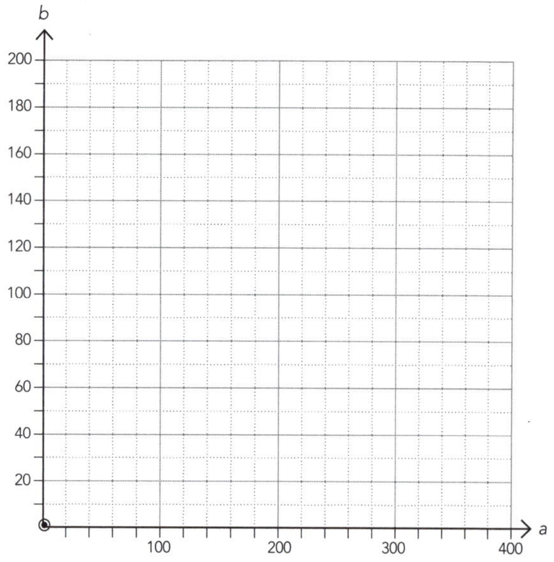

 ISBN: 9780170389396

3 Using the x and y intercepts

- Often you will need to draw the graph of an equation that is not in $y = mx + c$ form.
- If it is in the form **$ax + by = c$**, it is usually easier to use the **intercept-intercept** method in order to draw the graph.

Examples:

1 Draw the graph of $8x - 3y = 24$.

Step 1:
Make $x = 0$.
Then $8(0) - 3y = 24$
$\therefore y = \mathbf{-8}$
Plot the point (0, **-8**).

Step 2:
Make $y = 0$.
Then $8x - 3(0) = 24$
$\therefore x = \mathbf{3}$
Plot the point (**3**, 0).

Step 3:
Join the points.

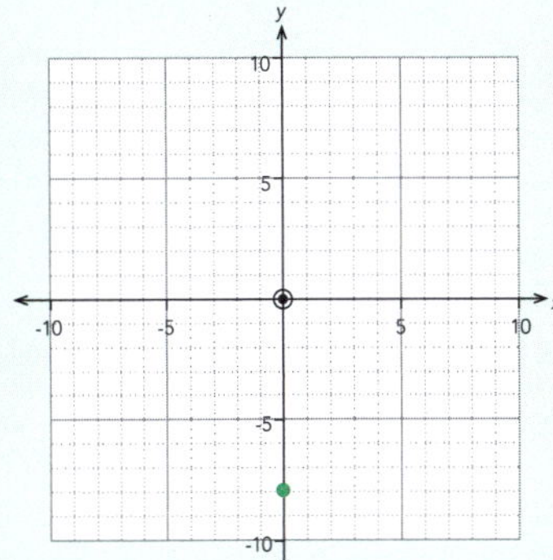

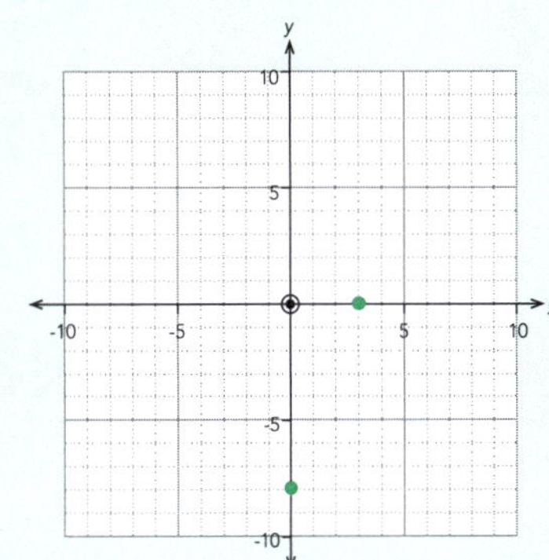

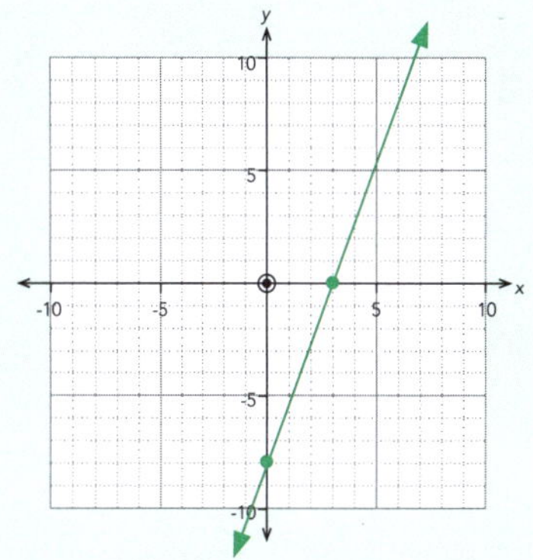

2 Draw the graph of $8x + 9y = 360$.

Step 1:
Make $x = 0$.
Then $8(0) + 9y = 360$
$\therefore y = \mathbf{40}$
Plot the point (0, **40**).

Step 2:
Make $y = 0$.
Then $8x + 9(0) = 360$
$\therefore x = \mathbf{45}$
Plot the point (**45**, 0).

Step 3:
Join the points.

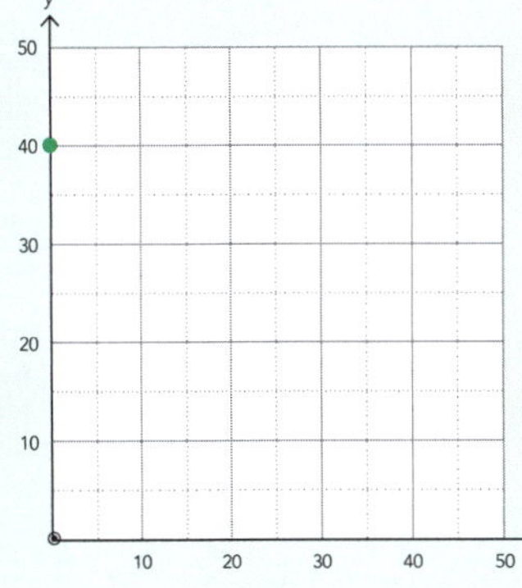

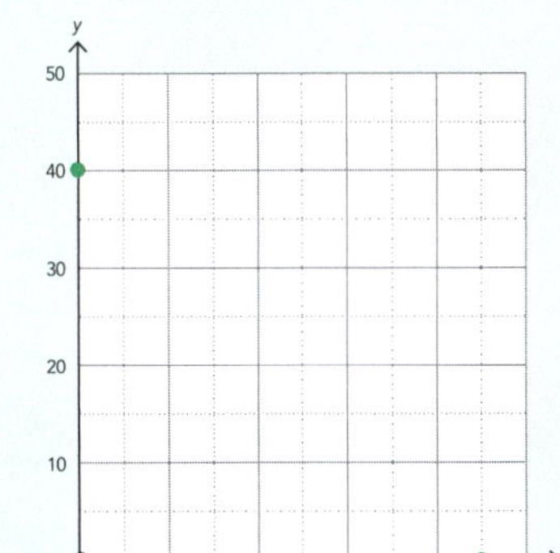

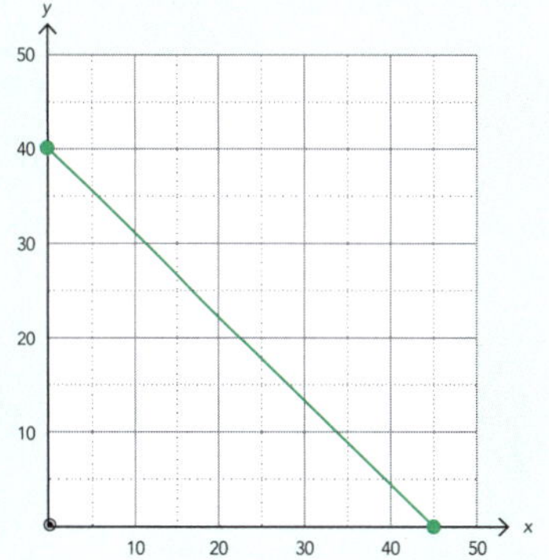

ISBN: 9780170389396

Draw the following lines.

1 $3x + 5y = 30$
$x = 0 \Rightarrow y =$ ____ $y = 0 \Rightarrow x =$ ____

2 $15x + 12y = 60$
$x = 0 \Rightarrow y =$ ____ $y = 0 \Rightarrow x =$ ____

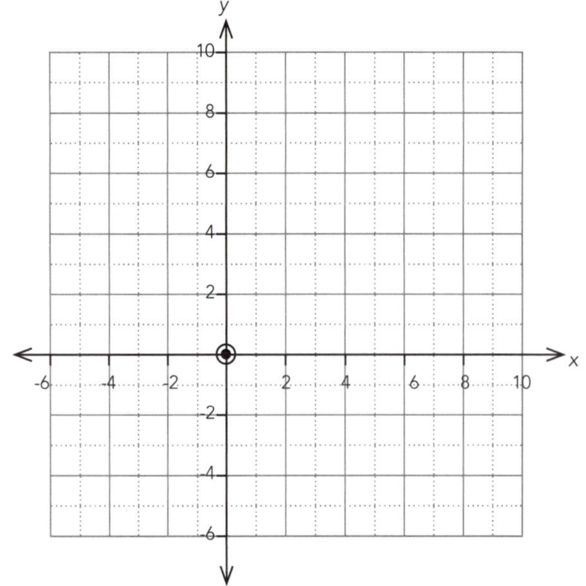

3 $3x + 2y = 120$

4 $3x + 5y = 180$

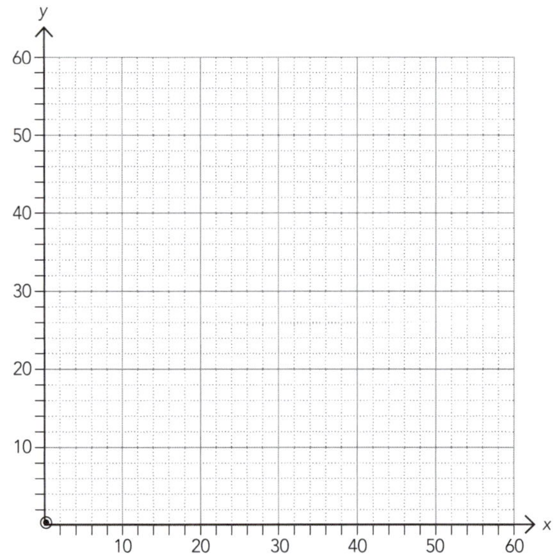

Hint: You may need to use combinations of methods sometimes.

5 $36x + 6y - 540 = 0$

6 $2y = 140 - 7x$

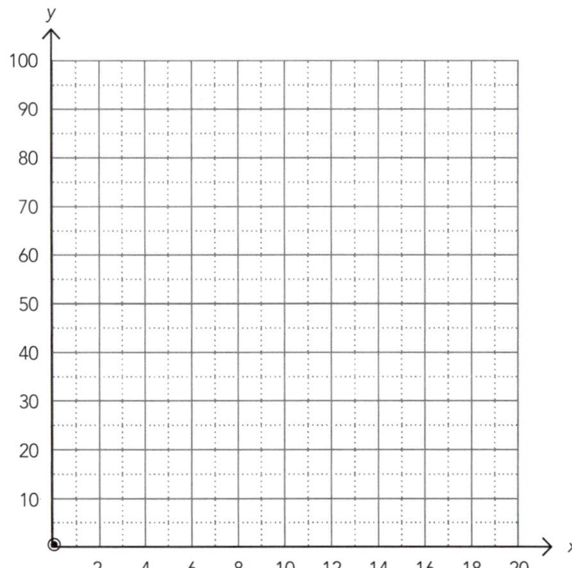

ISBN: 9780170389396

7 $100x + y = 600$

8 $100x + 2y - 1000 = 0$

9 $180 - 2x = 3y$

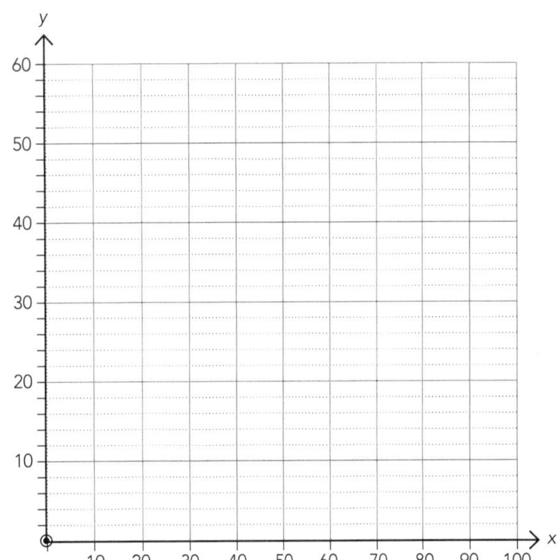

10 $4x = 240 - 5y$

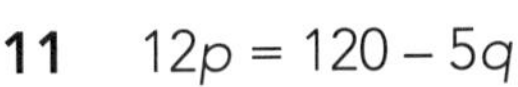
11 $12p = 120 - 5q$

12 $3q + 20p - 180 = 0$

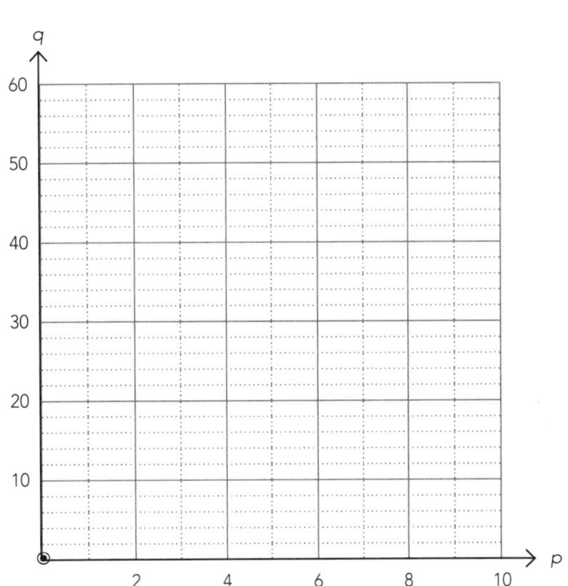

4 Horizontal and vertical lines

Be very careful when drawing these — it's easy to get them the wrong way around.

Examples:

1 Draw the line $x = 8$.

Step 1: Plot at least three points where $x = 8$, e.g. (8, 0), (8, 4), (8, 10).

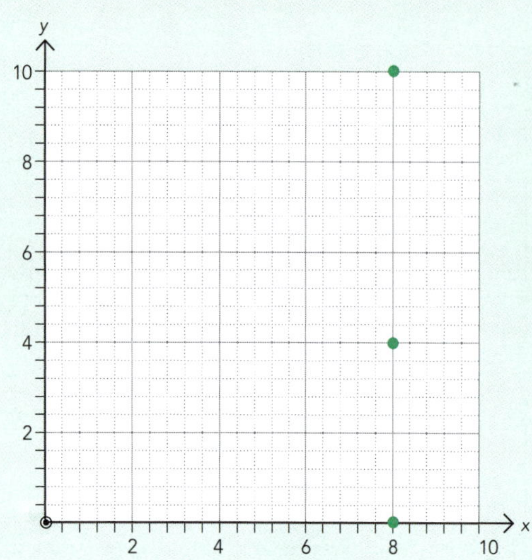

Step 2: Join the points.

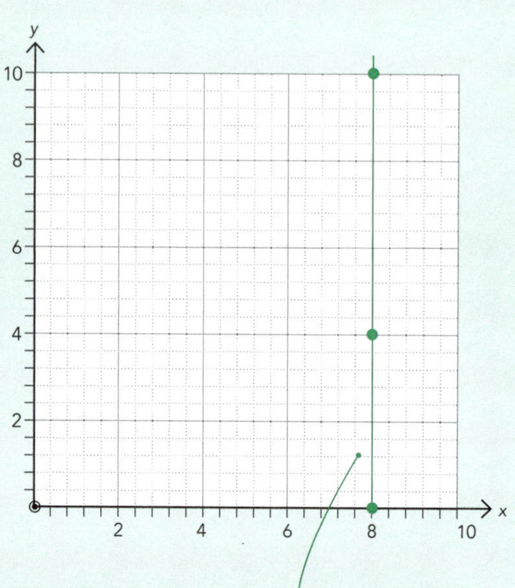

Notice that although this is $x = 8$, it is parallel with the y-axis.

2 Draw the line $b = 420$.

Step 1: Plot at least three points where $b = 420$, e.g. (0, 420), (5, 420), (10, 420).

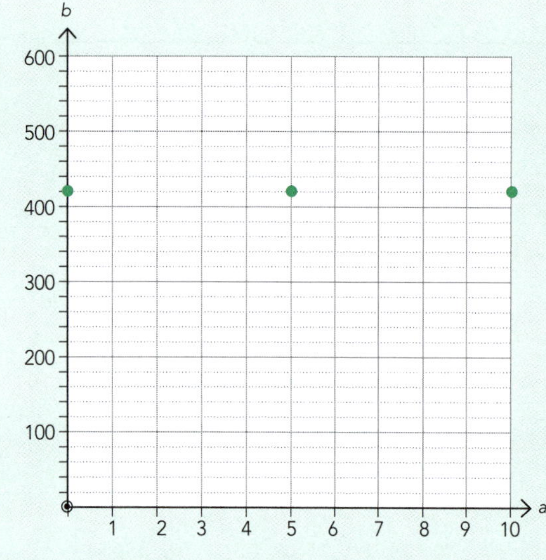

Step 2: Join the points.

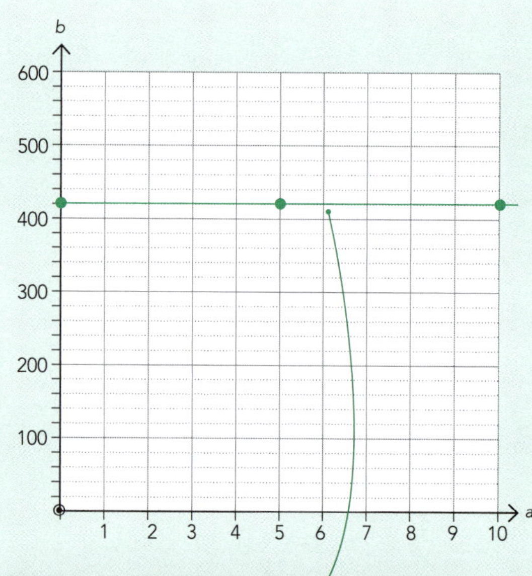

Again, although this is $b = 420$, it is parallel with the a-axis.

ISBN: 9780170389396

Draw the following lines.

1 $x = 5$ and $y = -2$

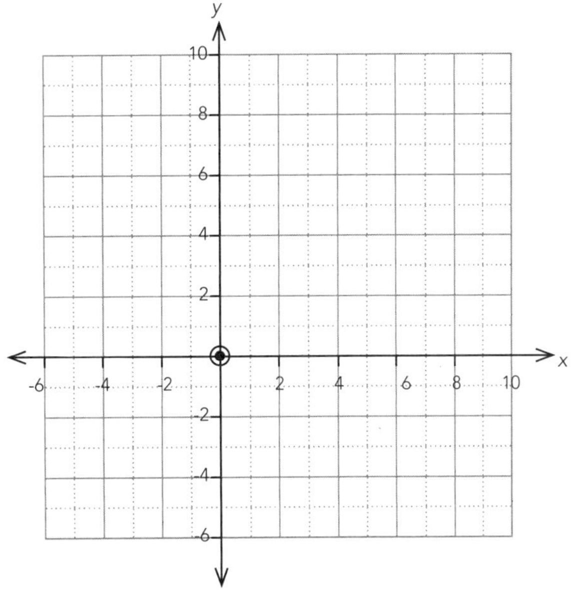

2 $y = 34$ and $x = 28$

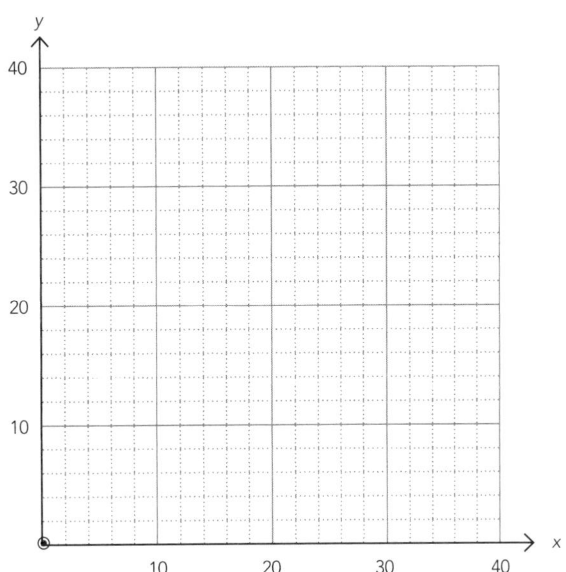

3 $p = 70$ and $q = 85$

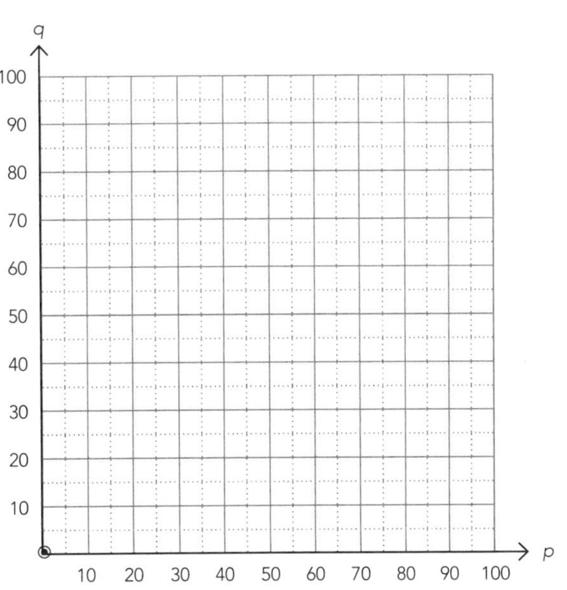

ISBN: 9780170389396

5 Drawing lines using your calculator

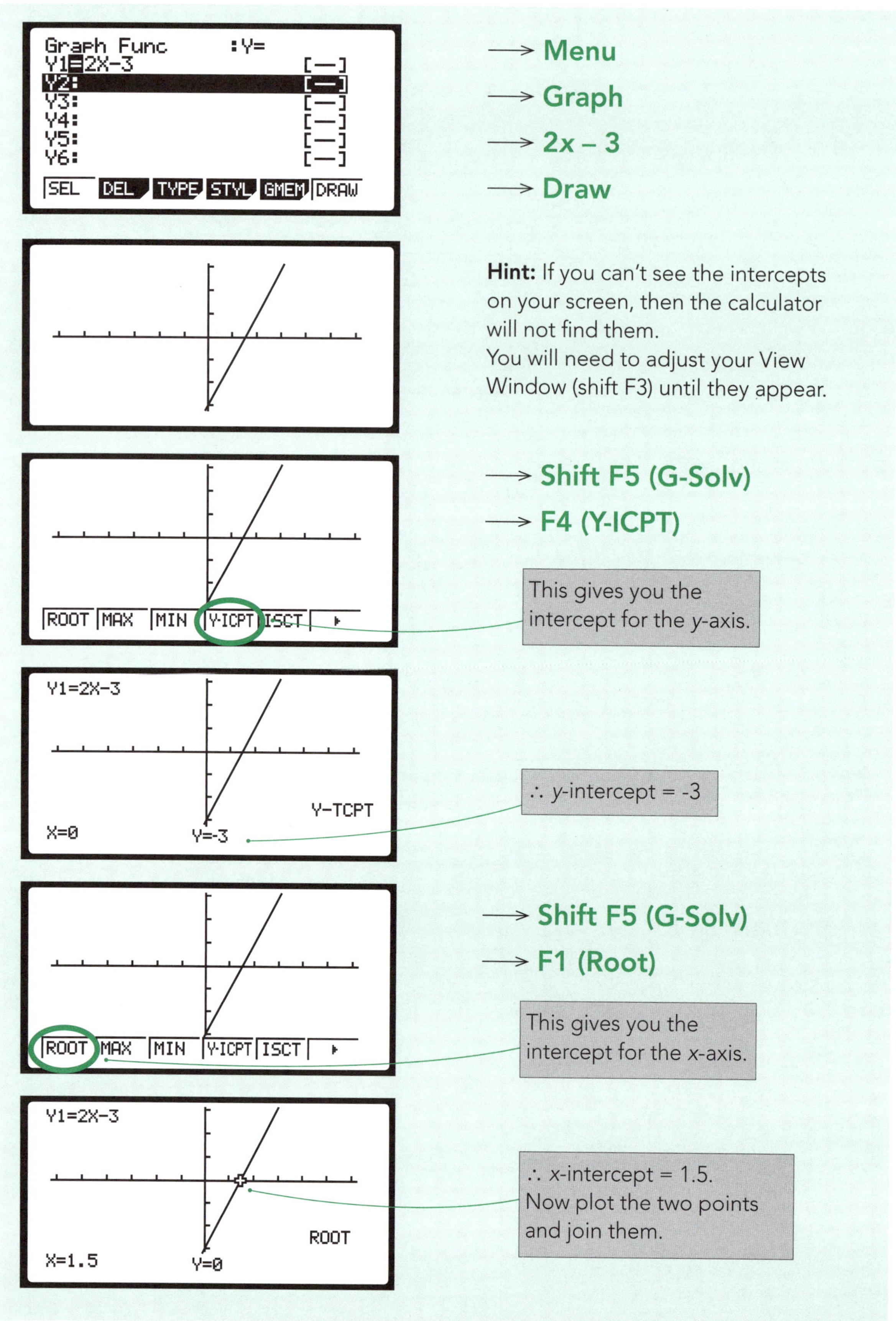

ISBN: 9780170389396

Test yourself

Use whatever method you wish to draw the following.

1 $y = x - 1$

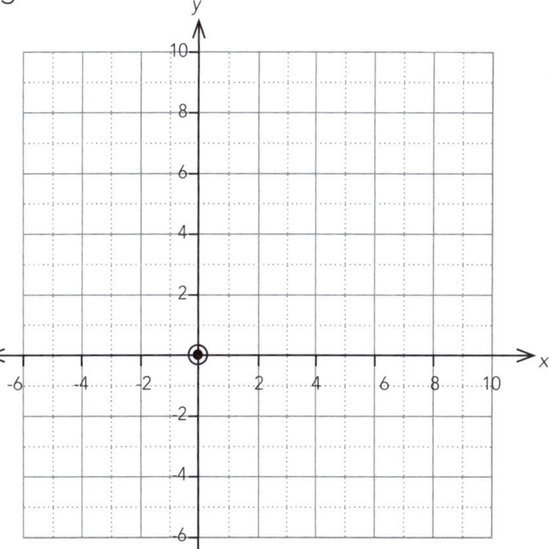

2 $y + 2x = 8$

3 $y - 2x = 6$

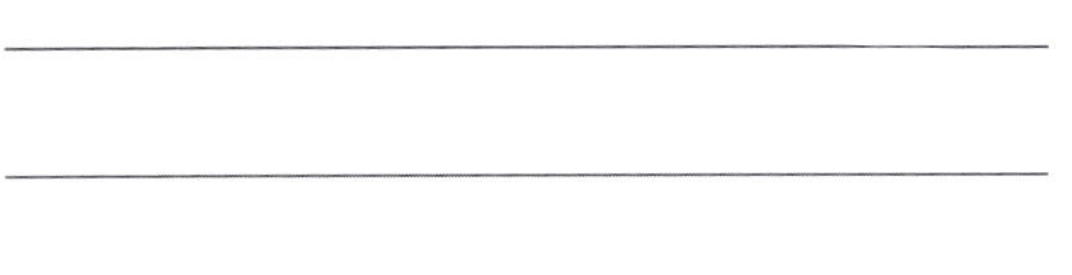

4 $y = 36 - 4x$

5 $y + 6x = 90$

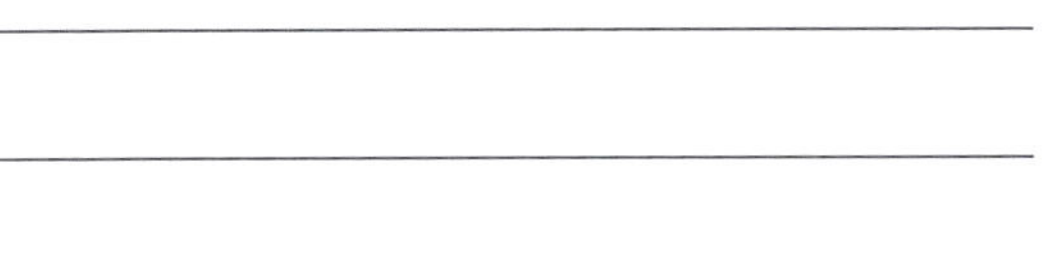

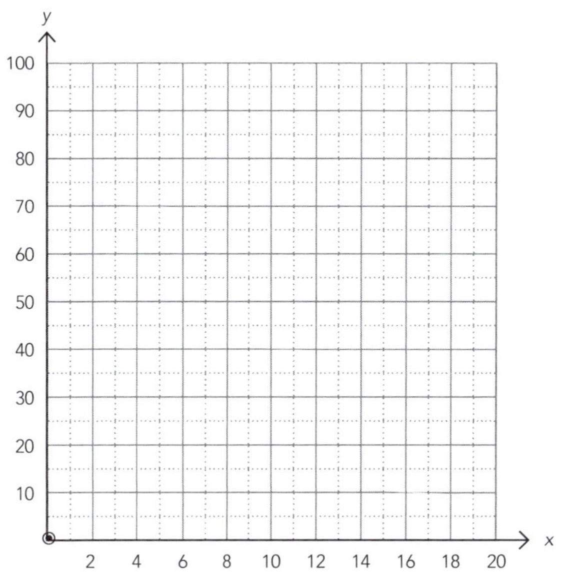

6 $5x - y + 20 = 0$

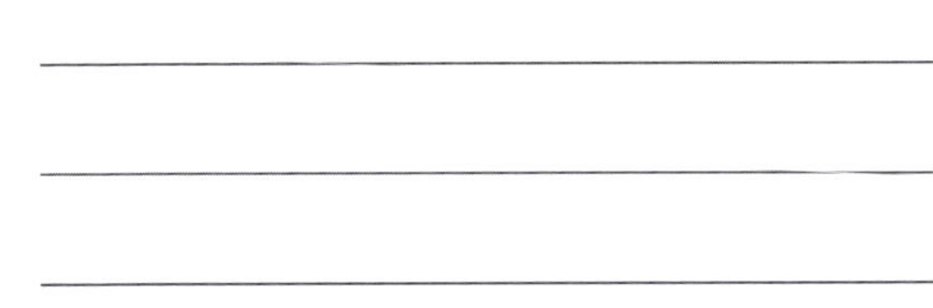

ISBN: 9780170389396

7 $5x + 6y = 3000$

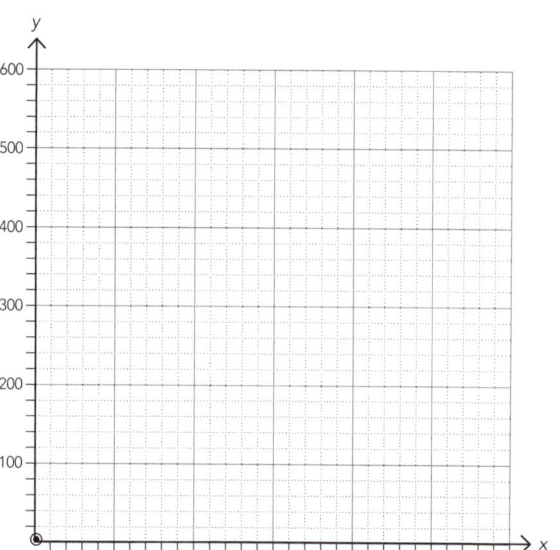

8 $2y = 360 + x$

9 $4y - x = 120$

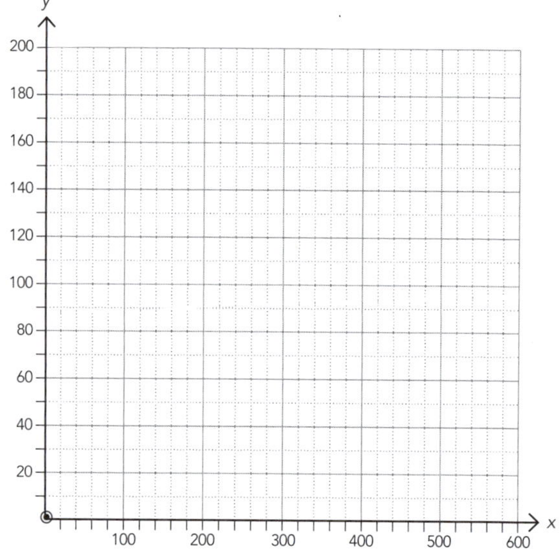

10 $2x + 5y - 900 = 0$

11 $2a = 5b$

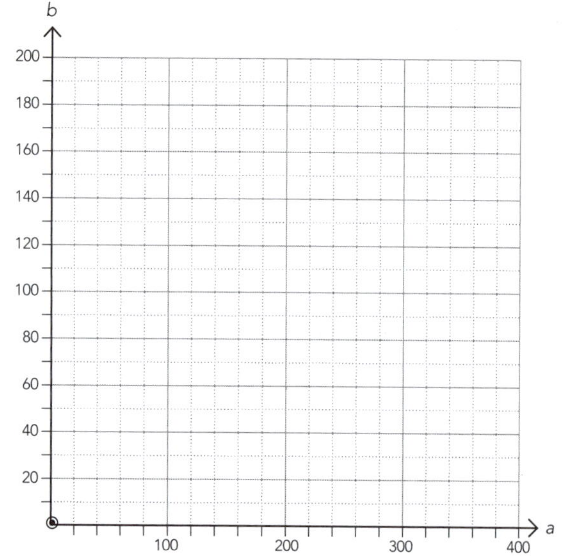

12 $1600 - 5a - 8b = 0$

ISBN: 9780170389396

Finding where lines meet (intersections)

1 By drawing the lines

Examples:

1 Find the coordinates of the point where the lines $y = 2x - 3$ and $4x + 5y = 20$ meet.

Step 1: Draw a table for $y = 2x - 3$.

x	2x – 3	y	(x, y)
0	2(0) – 3	-3	**(0, -3)**
2	2(2) – 3	1	**(2, 1)**
4	2(4) – 3	5	**(4, 5)**
6	2(6) – 3	9	**(6, 9)**

Plot and connect the points.

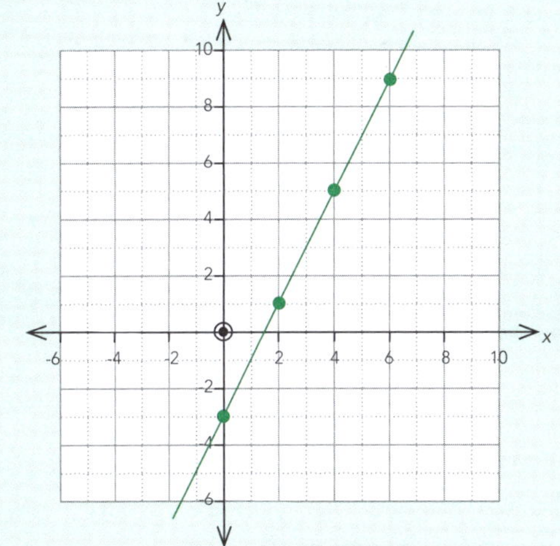

Step 2: Substitute 0 for x and y in $4x + 5y = 20$.

Let $x = 0$: $5y = 20$
$y = 4 \Rightarrow$ **(0, 4)**

Let $y = 0$: $4x = 20$
$x = 5 \Rightarrow$ **(5, 0)**

Plot and connect the points.

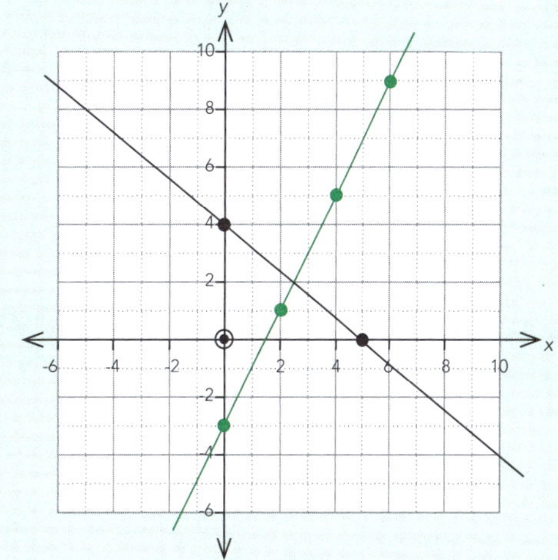

Step 3:

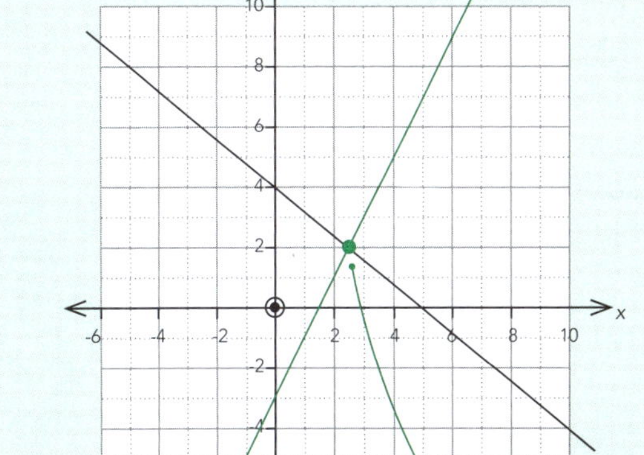

Probable solution: (2.5, 2).

You cannot be certain of this unless you:

a solve the equations simultaneously or use your calculator to confirm these values

or

b substitute these values into the two equations to check that they work.

ISBN: 9780170389396

2 Find the coordinates of the point where the lines $9x + 20y = 3600$ and $3x + 20y = 2400$ meet.

Step 1: Substitute 0 for x and y in $9x + 20y = 3600$.

Let $x = 0$: $20y = 3600$
$y = 180 \Rightarrow$ **(0, 180)**

Let $y = 0$: $9x = 3600$
$x = 400 \Rightarrow$ **(400, 0)**

Plot and connect the points.

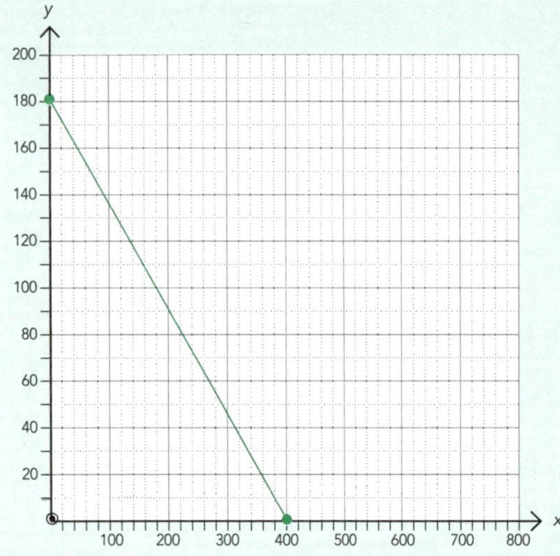

Step 2: Substitute 0 for x and y in $3x + 20y = 2400$.

Let $x = 0$: $20y = 2400$
$y = 120 \Rightarrow$ **(0, 120)**

Let $y = 0$: $3x = 2400$
$x = 800 \Rightarrow$ **(800, 0)**

Plot and connect the points.

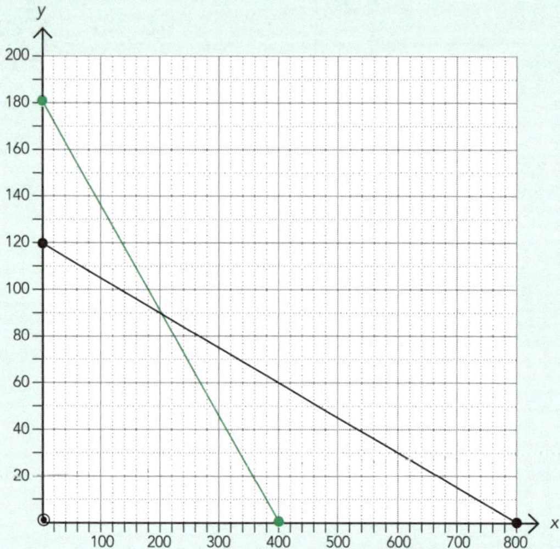

Step 3:

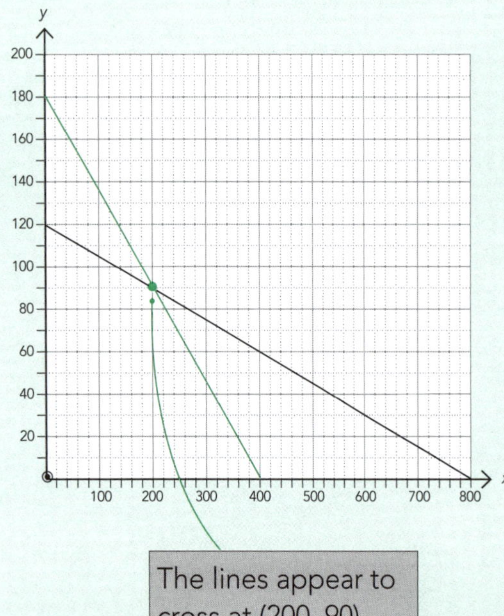

Probable solution: (200, 90).
Because of the scale, you **definitely cannot be certain** of this unless you:

a solve the equations simultaneously or use your calculator to confirm these values

or

b substitute these values into the two equations to check that they work.

ISBN: 9780170389396

Graph these pairs of equations and use the graph to estimate the point of intersection.

1 $5x + y = 40$
$y - 2x = 12$

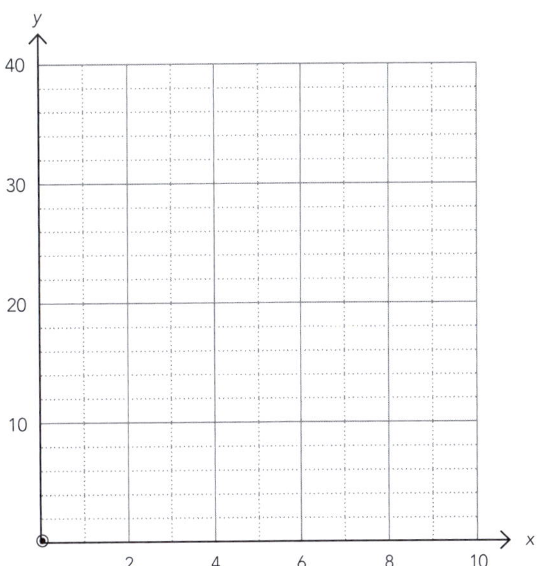

2 $6x + 10y = 3600$
$10y - 7x = 1000$

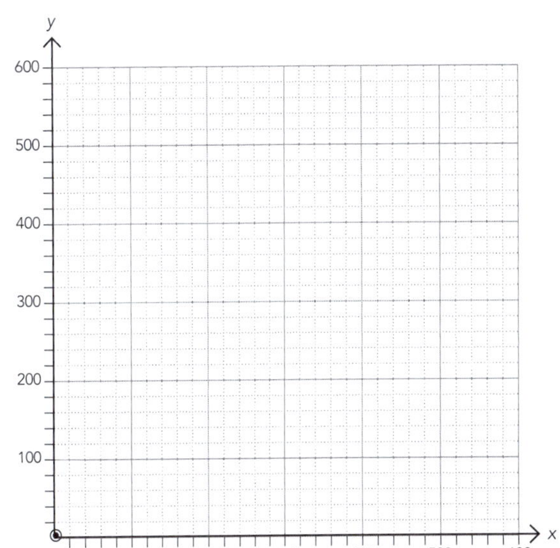

3 $10y - 3x = 200$
$20y + 3x = 4000$

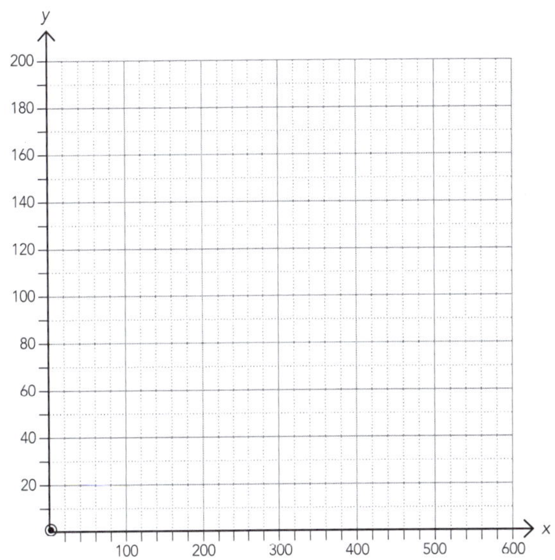

4 $5y - 2x = 1400$
$x + y = 420$

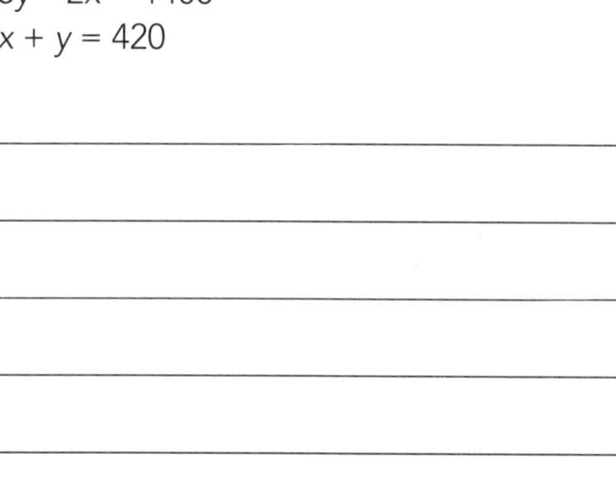

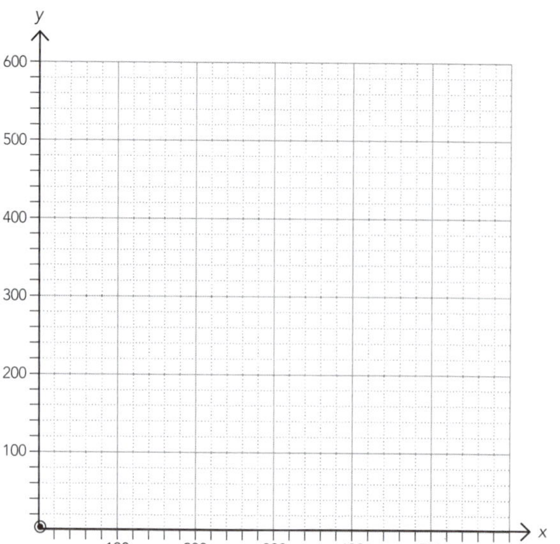

5 $5y - x = 125$
$30y - 2x - 1950 = 0$

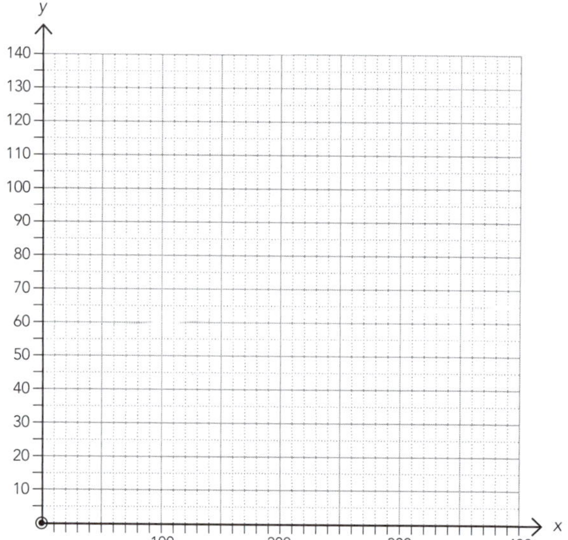

6 $y = 100 - 8x$
$4y + 7x = 200$

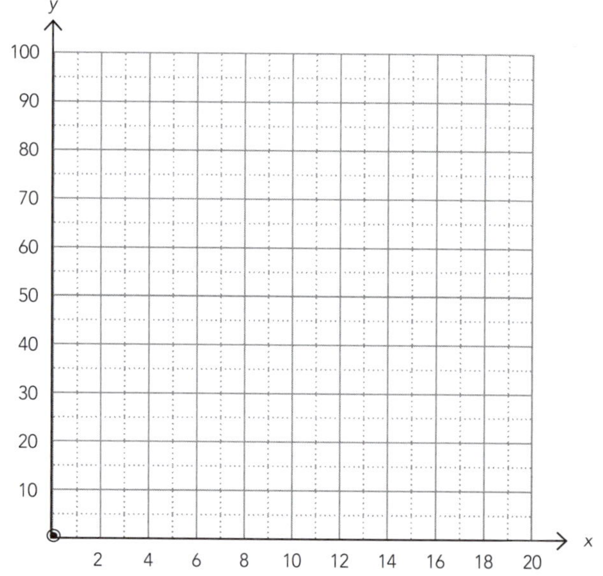

ISBN: 9780170389396

2 By solving the equations simultaneously

- 'Simultaneously' means '**at the same time**'.
- You need to solve equations simultaneously when you cannot read exact coordinates from your graph.
- There are **two methods** for doing this, and which one you use depends on how the equations are structured.

Substitution

- Substitution is easiest where one of the equations is expressed as **$x =$ …** or **$y =$ …**.
- As the name suggests, you substitute the **$x =$ …** into the other equation.
- You should number each equation, and say what you are doing at each step.

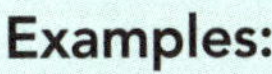

Examples:

1 Solve the equations $y = 2x - 3$ and $4x + 5y = 20$.

$\mathbf{y = 2x - 3}$ ①

$4x + 5y = 20$ ②

Substitute ① into ②: $4x + 5(\mathbf{2x - 3}) = 20$

$4x + 10x - 15 = 20$

$14x - 15 = 20$

$\mathbf{x = 2.5}$

Substitute for x in ①: $y = 2(2.5) - 3$

$\mathbf{y = 2}$ ∴ **The lines meet at (2.5, 2).**

> Number the equations.

> From equation ① we know that **y** and **$2x - 3$** are equal.

2 Solve the equations $18x + 11y = 381$ and $y = 21x - 237$.

$18x + 11y = 381$ ①

$y = \mathbf{21x - 237}$ ②

Substitute ② into ①: $18x + 11(\mathbf{21x - 237}) = 381$

$18x + 231x - 2607 = 381$

$249x = 2988$

$\mathbf{x = 12}$

Substitute for x in ②: $y = 21(12) - 237$

$\mathbf{y = 15}$ ∴ **The lines meet at (12, 15).**

> Select ② because it has the form $y =$ …

> Select whichever equation is easier for substitution.

Try for yourself.

1

$x = 2y$

$4x + y = 9$

____

____

____

____

____

You should get (2, 1).

2

$y = x - 1$

$5x - 3y = 9$

____

____

____

____

____

You should get (3, 2).

ISBN: 9780170389396

Elimination

- Elimination is easiest when the two equations have the **same structure**.

For example:

$$\begin{aligned} 3x &+ y = 11 \\ 2x &+ 3y = 12 \end{aligned}$$

x terms | y terms | = signs | constants

- You solve these by multiplying one or both equations by a constant in order to make either the x terms or the y terms the **same size** but with **different signs**.
- Then **add** the two equations to produce an equation with one variable only.
- You should number each equation, and say what you are doing at each step.

Examples:

1 Solve the equations $3x + y = 11$ and $2x + 3y = 12$.

$3x + y = 11$ ①

$2x + 3y = 12$ ②

Multiply ① by -3: $-9x - 3y = -33$ ③

Call this ③ because it is a new equation.

Add ② and ③: $-7x = -21$

$\therefore$ **$x = 3$**

Substitute for x in ①: $3(3) + y = 11$

$\therefore$ **$y = 2$** $\quad\therefore$ **The lines meet at (3, 2).**

2 Solve the equations $4x + 3y = 43$ and $2x + 5y = 53$.

$4x + 3y = 43$ ①

$2x + 5y = 53$ ②

Multiply ① by -5 $\quad -20x - 15y = -215$ ③

Multiply ② by 3: $\quad 6x + 15y = 159$ ④

You need to multiply **both** equations by a constant.

Add ③ and ④: $-14x = -56$

$\therefore$ **$x = 4$**

Substitute for x in ②: $2(4) + 5y = 53$

$5y = 45$

$\therefore$ **$y = 9$** $\quad\therefore$ **The lines meet at (4, 9).**

Try for yourself.

1

$2x + 3y = 14$

$5x + 3y = 26$

You should get (4, 2).

2

$5x + 3y = 24$

$4x - y = 26$

You should get (6, -2).

ISBN: 9780170389396

3 Finding intersections using your calculator

Find the coordinates of the point where the lines $y = 2x + 1$ and $2y + x = 12$ meet.

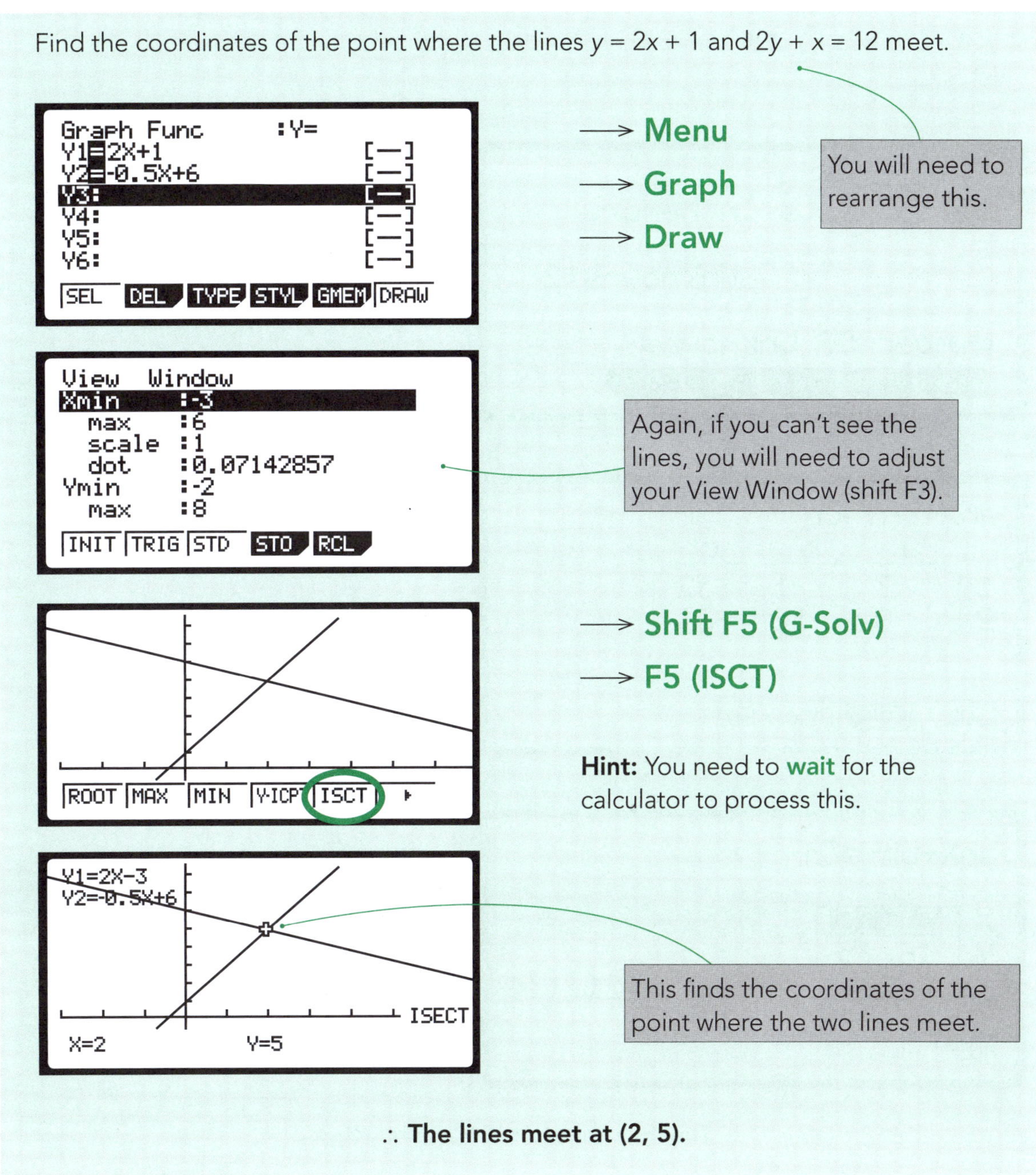

∴ **The lines meet at (2, 5).**

Try for yourself.

1

$$x = 2y$$
$$4x + y = 9$$

You should get (2, 1).

2

$$y = x - 1$$
$$5x - 3y = 9$$

You should get (3, 2).

ISBN: 9780170389396

4 Solving equations simultaneously using your calculator

Find the coordinates of the point where the lines $y = 7x - 132$ and $8x - 12y + 31 = 0$ meet.

```
Simultaneous
Data Exists In Memory

              Unknowns:2

Number Of Unknowns?
 2   3   4   5   6
```

→ **Menu**
→ **Equa**
→ **F1: Simultaneous**
→ **F1: 2 Unknowns**

```
anX+bnY=Cn
    a      b      c
1 [ 7     -1    132 ]
2 [ 8    -12    -31 ]

SOLV DEL CLR EDIT
```

This shows you the format in which your equation must be.

Example:

$y = 7x - 132$ ①
$8x - 12y + 31 = 0$ ②

Rearrange ⇒ $ax + by = c$

$7x - \quad y = 132$ ①
$8x - 12y = -31$ ②

```
anX+bnY=Cn

X [ 21.25 ]
Y [ 16.75 ]
                  21⌟1⌟4

REPT
```

→ **F1: Solve**

∴ The lines meet at (21.25, 16.75).

Try for yourself.

1

$25x + 10y = 0$
$-x + 6y = 32$

You should get (-2, 5).

2

$3x = 9 + 4y$
$-2x + 3y = 10$

You should get (67, 48).

 ISBN: 9780170389396

Test yourself

Use an appropriate method to solve the following.

1
$$y = -3x$$
$$x + y = 6$$

2
$$x - y = -7$$
$$4x + 3y = 14$$

3
$$y = x + 9$$
$$13x + 2y = 183$$

4
$$12x + 5y = 139$$
$$7x - y = 38$$

5
$$x = 114 - 15y$$
$$20x + 23y = 341$$

6
$$9x - 4y = 90$$
$$7x + 5y = 289$$

7
$$5x + y = 40$$
$$y - 2x = 12$$

8
$$y = 100 - 8x$$
$$4y + 7x = 200$$

9
$$x = 6y + 20$$
$$5x + 10y - 160 = 0$$

10
$$y = 7x - 132$$
$$8x - 12y + 31 = 0$$

ISBN: 9780170389396

11

$y = x + 9$
$13x + 2y = 183$

12

$15x + 8y = 519$
$x = 2y - 11$

13

$x = 114 - 15y$
$20x + 23y = 341$

14

$x = 124 - 3y$
$7x + 5y = 356$

15

$x = 5y - 75$
$13y - 71 = 15x$

16

$18x - 11y - 15 = 0$
$y = 4x - 12$

17

$3x = 20 + y$
$4y + 7x = 129$

18

$4y - x = 37$
$y + 6x = 73$

19

$5x = 708 - 17y$
$13x - 298 = 9y$

20

$6y + 264 - 15x = 0$
$5x = 577 - 4y$

 ISBN: 9780170389396

Inequations

Understanding the words

Before you can draw the graphs, you need to be able to convert sentences into inequations that contain the symbol $<$, $>$, $\leq$ or $\geq$, so you need to know the **words** that represent each of the symbols.

When drawing graphs of inequations:

1 If values along the line **are** included ($\leq$ or $\geq$), use a solid line: ————

If values along the line **are not** included ($<$ or $>$), use a dashed line: -----------

2 Shade the area that is **not** true.

Note: From here on we will consider only **positive** values for x and y.

Words	The area containing values that x can take is left unshaded	Notation
x is **more than** 5 x is **greater than** 5 x is **over** 5 x **exceeds** 5	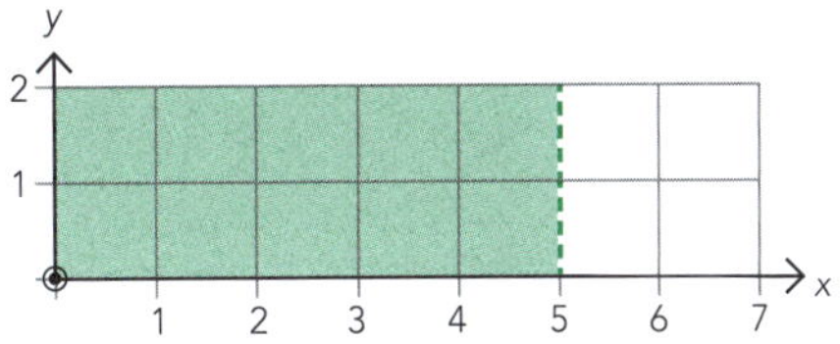	$x > 5$
x is **less than** 5 x is **fewer than** 5 x **is under** 5	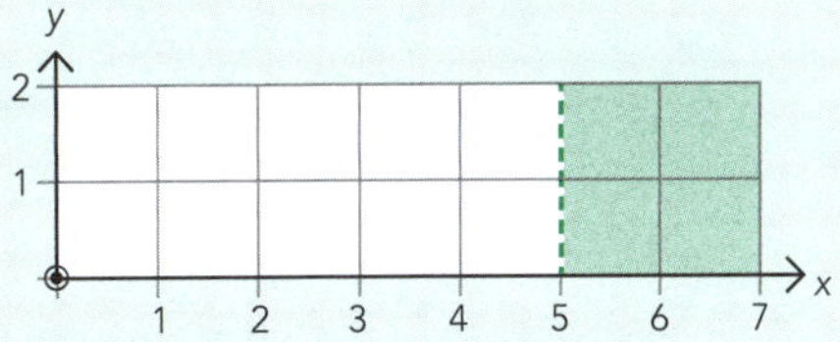	$x < 5$
x is **greater than or equal to** 5 x is **at least** 5 x is **no(t) less than** 5 x is **5 or more** x is a **minimum** of 5	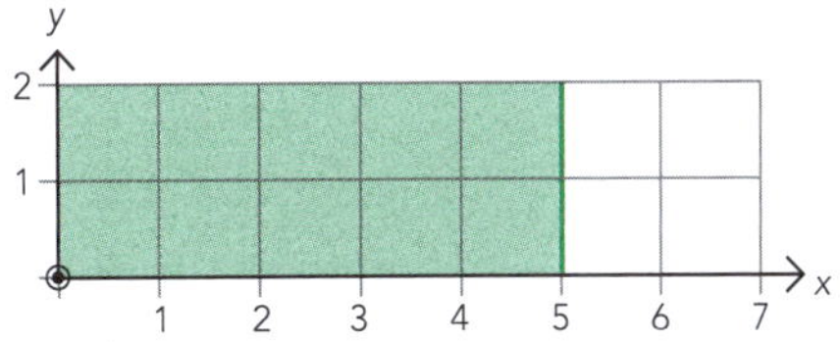	$x \geq 5$
x is **less than or equal to** 5 x is 5 **or less** x is **no(t) more than** 5 x is **at most** 5 x is a **maximum** of 5	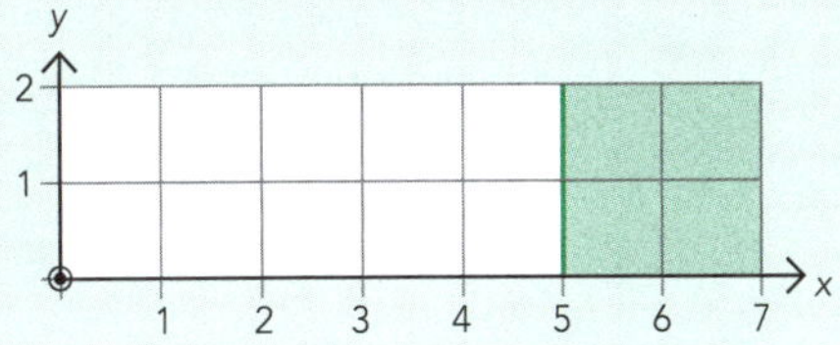	$x \leq 5$
x is **between** 1 and 4	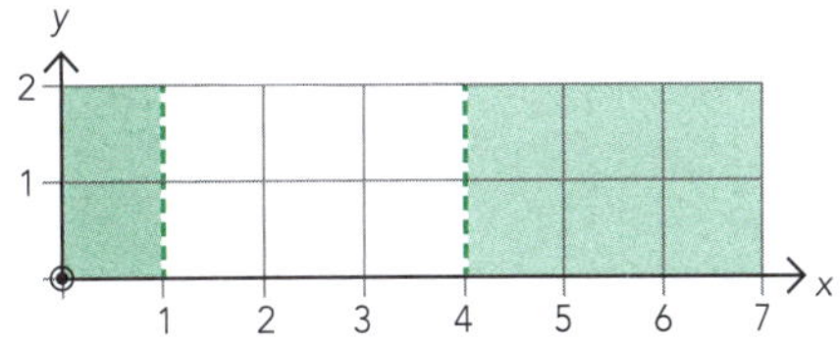	$1 < x < 4$
x is **between** 1 and 4 **inclusive**	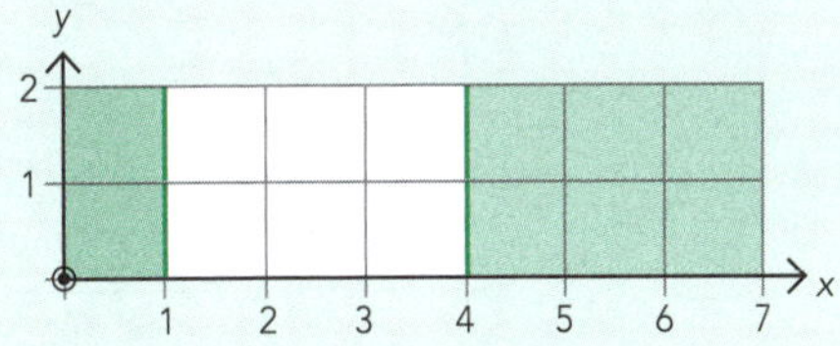	$1 \leq x \leq 4$

1 Complete the graphs and write the notation represented by each of the following expressions.

Words	Shade the areas containing values that x cannot take	Notation
x is greater than 2		$x > 2$
x is between 3 and 7		
x is less than 5		
x is at least 6		
x is between 1 and 4 inclusive		
x is greater than or equal to 1		
x is 5 or less		

ISBN: 9780170389396

Words	Shade the areas containing values that x cannot take	Notation
x is under 7	y: 1, 2; x: 1, 2, 3, 4, 5, 6, 7, 8	
x is over 6	y: 1, 2; x: 1, 2, 3, 4, 5, 6, 7, 8	
x is 2 or more	y: 1, 2; x: 1, 2, 3, 4, 5, 6, 7, 8	
x is more than 5	y: 1, 2; x: 1, 2, 3, 4, 5, 6, 7, 8	
x is not more than 3	y: 1, 2; x: 1, 2, 3, 4, 5, 6, 7, 8	
x exceeds 4	y: 1, 2; x: 1, 2, 3, 4, 5, 6, 7, 8	
x is not less than 7	y: 1, 2; x: 1, 2, 3, 4, 5, 6, 7, 8	

ISBN: 9780170389396

2 Complete the graphs and write words represented by the following notations.

Notation	Shade the areas containing values that *x* cannot take	Words
$x < 6$	y: 2, 1; x: 1 2 3 4 5 6 7 8	**x is less than 6**
$x \geq 2$	y: 2, 1; x: 1 2 3 4 5 6 7 8	
$1 < x < 4$	y: 2, 1; x: 1 2 3 4 5 6 7 8	
$x > 3$	y: 2, 1; x: 1 2 3 4 5 6 7 8	
$x \leq 4$	y: 2, 1; x: 1 2 3 4 5 6 7 8	
$2 \leq x \leq 6$	y: 2, 1; x: 1 2 3 4 5 6 7 8	

  ISBN: 9780170389396

Rearranging inequations

In order to graph inequations, you will often need to **rearrange** the terms.

Rules for rearrangement of inequations:

1 There should be only **one** $<$, $>$, $\leq$ or $\geq$ per line.

2 When you want to move a term to the other side, perform the **opposite** operation.

3 You can do anything you like as long as you do the **same** to both sides, **BUT** if you need to multiply or divide by a **negative number**, you must **reverse** the signs.

Conventions for $ax + by = c$:

1 No fractions for **a**, **b** or **c**.

2 **a** is positive.

Trick: If you want to change the signs of everything, **multiply or divide both sides by -1**.

Examples:

1 Convert $y < x$ into $ax + by = c$ form.

$y < x$ (– x)

$y - x < 0$ (**x by -1**)

$-y + x > 0$

$x - y > 0$

2 Convert $20 - y + 6x \geq 0$ into $y = mx + c$ form.

$20 - y + 6x \geq 0$ (– 20, – 6x)

$-y \geq -20 - 6x$ (**x by -1**)

$y \leq 20 + 6x$

$y \leq 6x + 20$

Multiplying by a **negative** number ⇒ reverse the sign.

3 Convert $\frac{1}{2}x - 5 \geq y$ into $ax + by = c$ form.

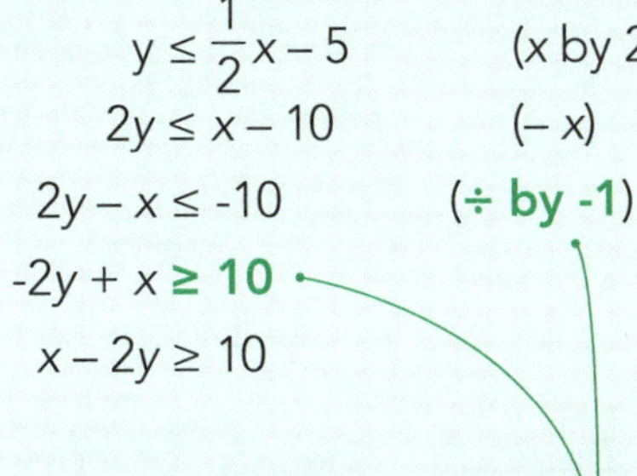

$y \leq \frac{1}{2}x - 5$ (x by 2)

$2y \leq x - 10$ (– x)

$2y - x \leq -10$ (**÷ by -1**)

$-2y + x \geq 10$

$x - 2y \geq 10$

4 Convert $3x - 2y > 18$ into $y = mx + c$ form.

$3x - 2y > 18$ (– 3x)

$-2y > 18 - 3x$ (**÷ by -2**)

$y < -9 + 1.5x$

$y < 1.5x - 9$

Dividing by a **negative** number ⇒ reverse the sign.

ISBN: 9780170389396

Convert the following inequations into the form required.

1 $5x + 2y \geq 30$ into $y = mx + c$ form

2 $y > \frac{5}{2}x + 1$ into $ax + by = c$ form

3 $x - y \leq 0$ into $y = mx + c$ form

4 $12 + y - 4x \geq 0$ into $ax + by = c$ form

5 $6x - 2y + 20 > 0$ into $y = mx + c$ form

6 $2y - 36 < 9x$ into $ax + by = c$ form

7 $x < 5y + 6$ into $y = mx + c$ form

8 $y \leq 4x - 11$ into $ax + by = c$ form

9 $102 - 24x - 6y > 0$ into $y = mx + c$ form

10 $20y + 100 \geq 95x$ into $ax + by = c$ form

ISBN: 9780170389396

Drawing inequations

- When linear programming, you almost always deal with **inequations**.
- You will need to draw the inequations on a **graph**.

Remember the rules for drawing inequations:

1 Lines: if values along the line **are** included (**≤** or **≥**), use a solid line; if values along the line **are not** included (**<** or **>**), use a dashed line.

2 Once you have drawn the line, you need to decide which side of it is true.

3 Shade the area that is **not** true.

Examples:

1 Draw the graph of $y < -2x + 36$.

Step 1: Plot a **dashed** line to show

$y = -2x + 36$

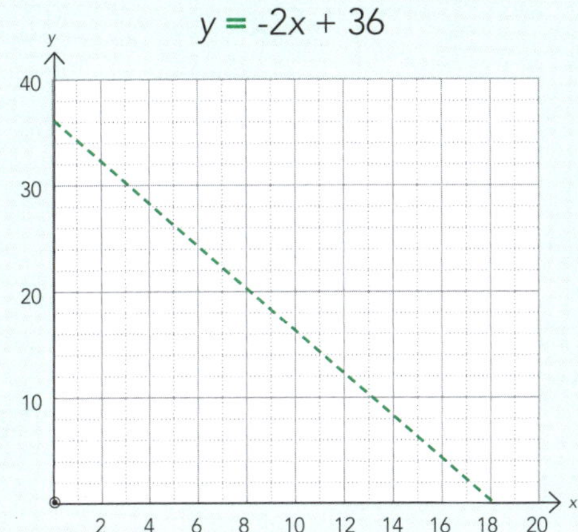

Step 2: Decide which side of the line is true.

a Select a point that is clearly on one side of the line. If appropriate, (0, 0) is an easy point to use.

b Substitute your chosen point into the equation:

$$0 < -2(0) + 36$$
$$0 < 36$$

True statement ⇒ equation **is** true on the side with (0, 0) ∴ true **below** the line.

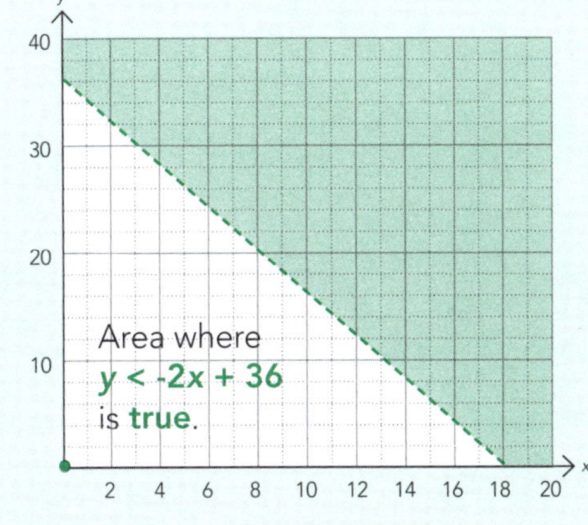

2 Draw the graph of $y \geq 2x$.

Step 1: Plot a **solid** line to show

$y = 2x$

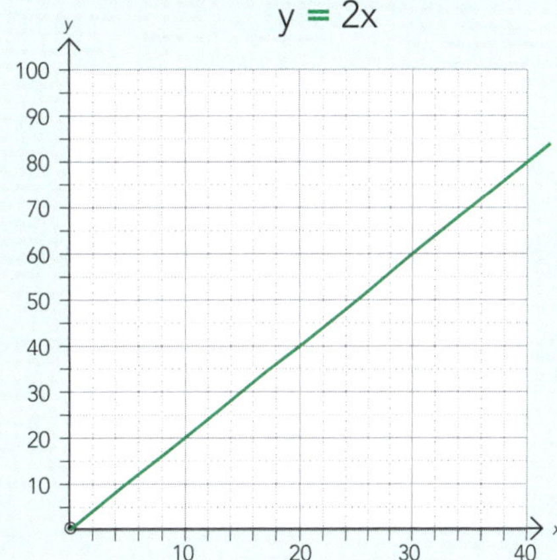

Step 2: Decide which side of the line is true.

a Select a point that is clearly on one side of the line. (0, 0) is on the line, so use (10, 0).

b Substitute your chosen point into the equation:

$$0 \geq 2(10)$$
$$0 \geq 20$$

Untrue statement ⇒ equation **is not** true on the side with (10, 0) ∴ true **above** the line.

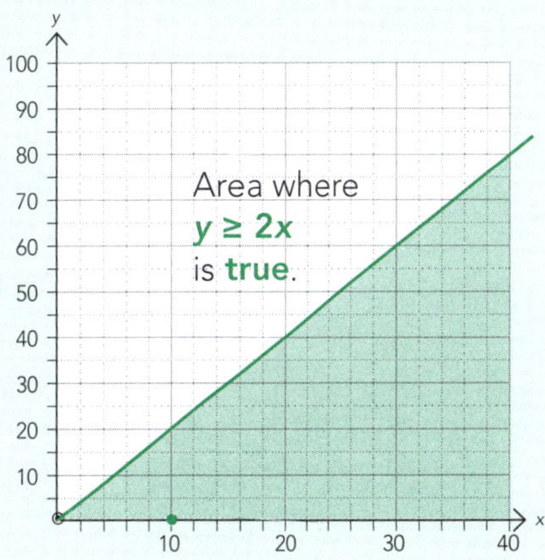

ISBN: 9780170389396

1 Match each of the following inequations with its correct graph.

a $y \geq 4$ ______

b $x \geq 4$ ______

c $y < 4$ ______

d $y > 4$ ______

e $x < 4$ ______

f $x \leq 4$ ______

g $x > 4$ ______

h $y \leq 4$ ______

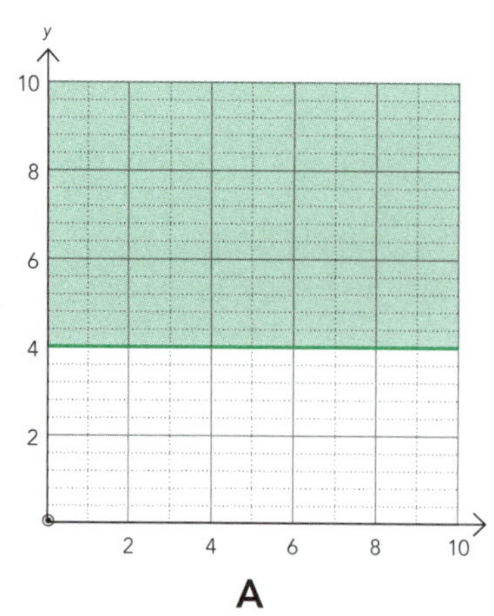

A

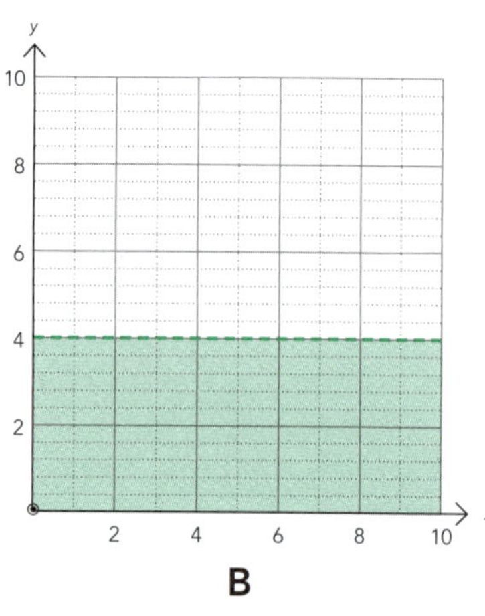

B

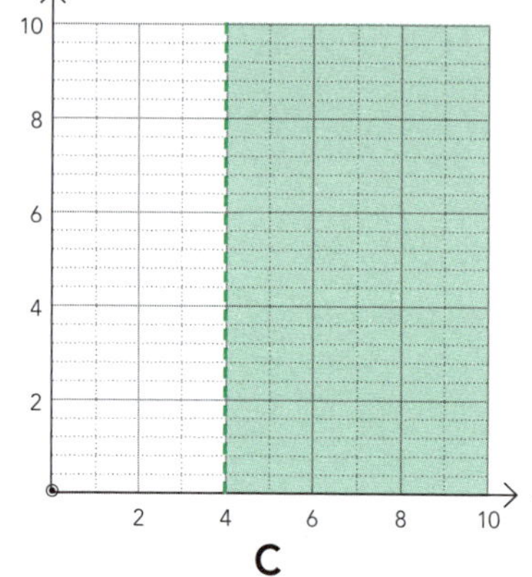

C

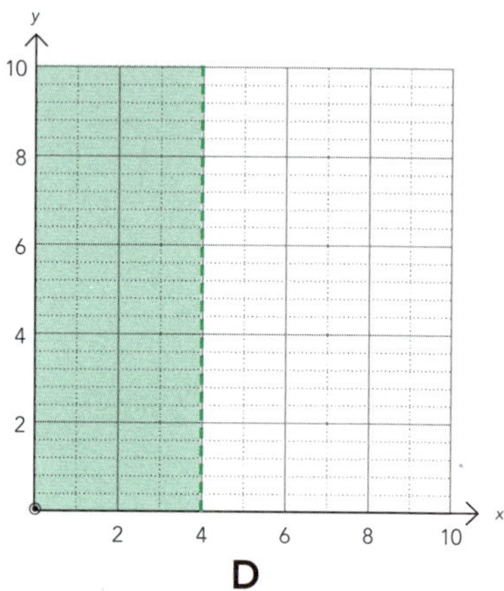

D

E

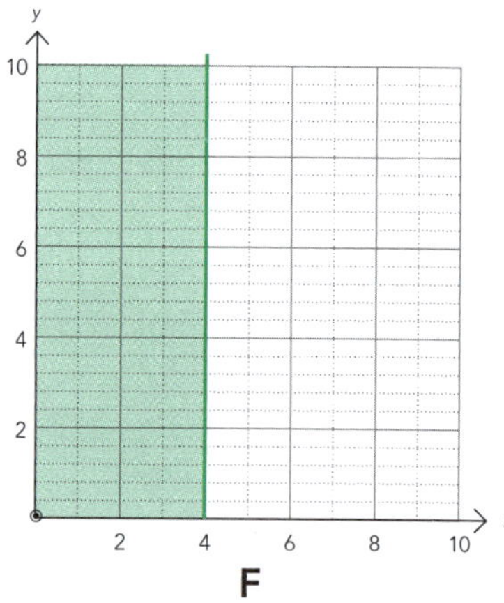

F

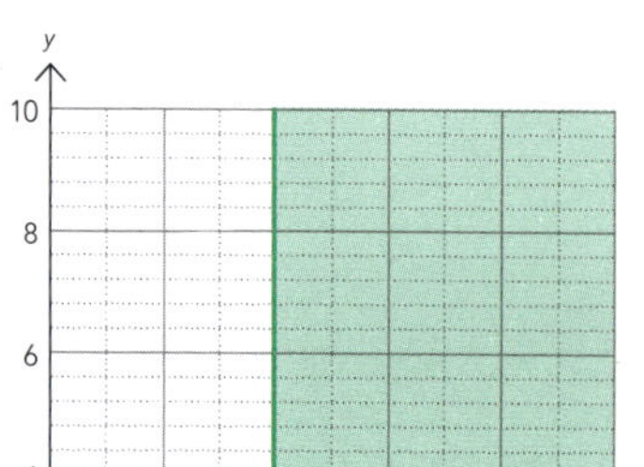

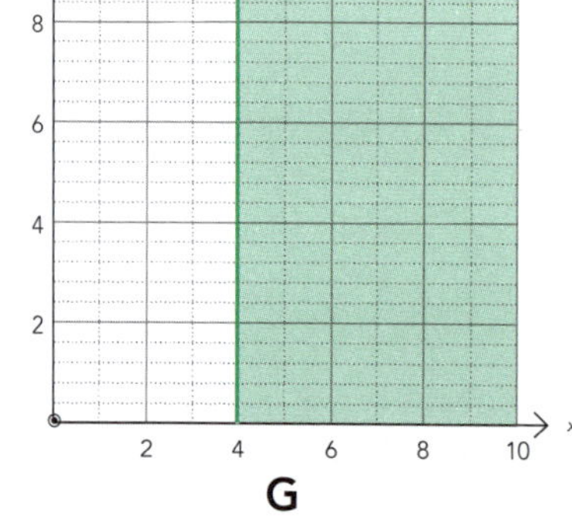

G

H

 ISBN: 9780170389396

2 Shade the area that is not true for the following inequations.

a $x > 2$

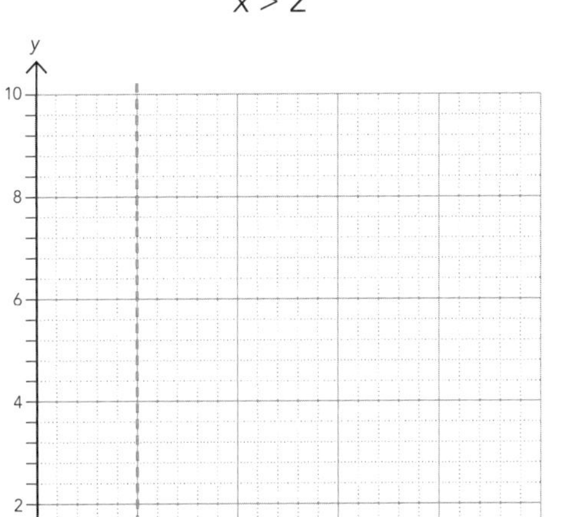

b $y \le 8$

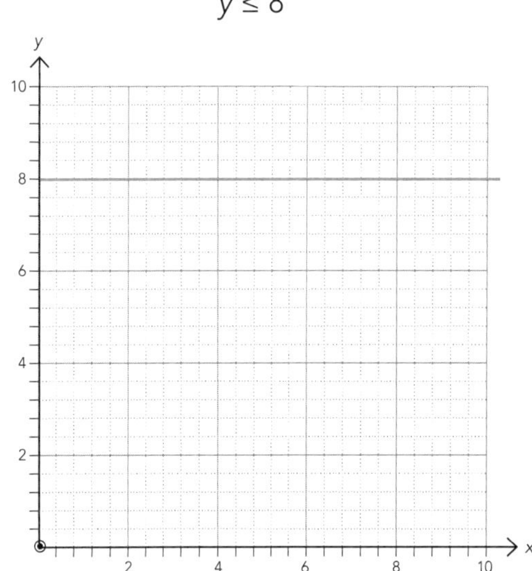

c $y \ge 9$

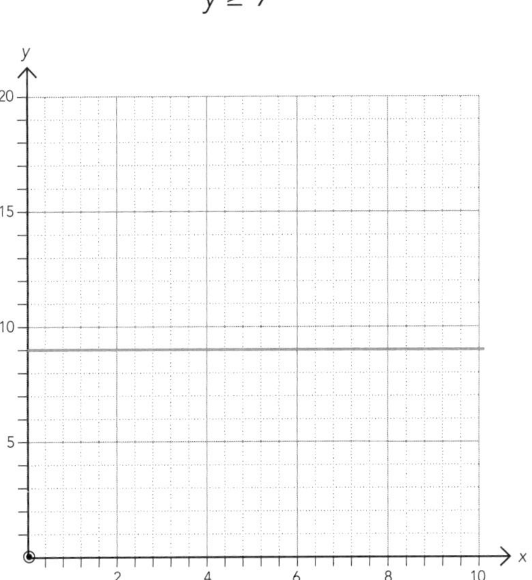

d $x \le 32$

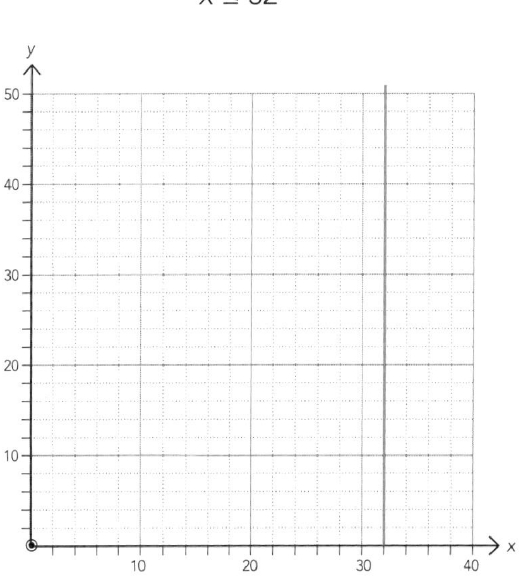

e $x \ge 45$

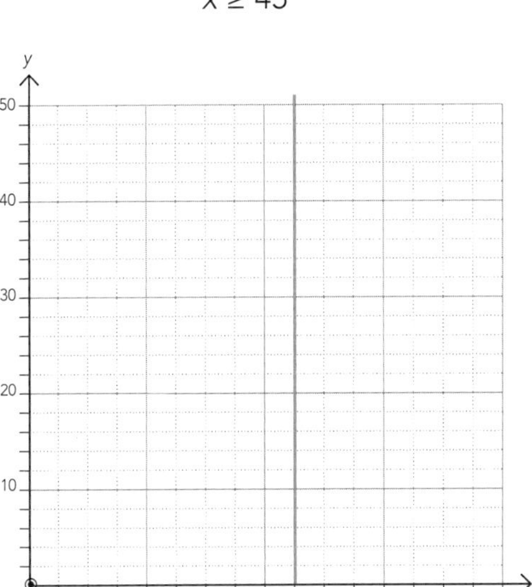

f $y > 65$

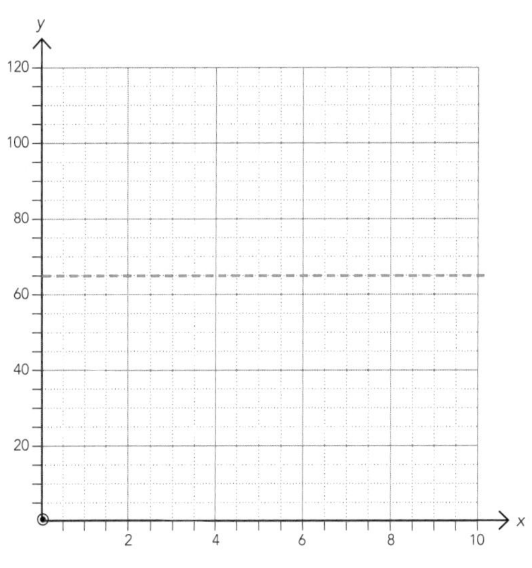

ISBN: 9780170389396

3 Draw graphs for the following inequations, and shade the area that is not true.

a $y \geq 7$

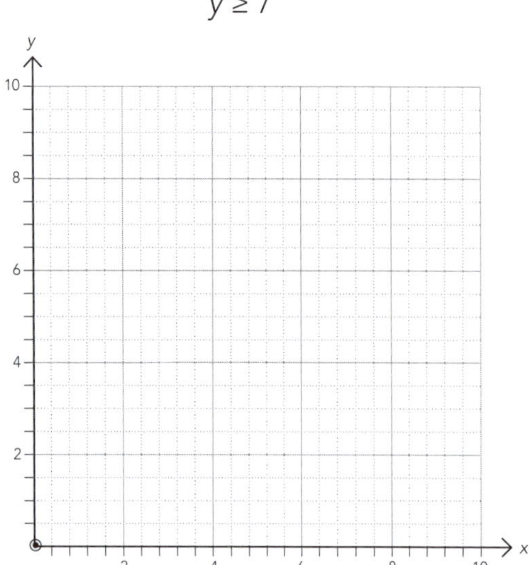

b $x < 6$

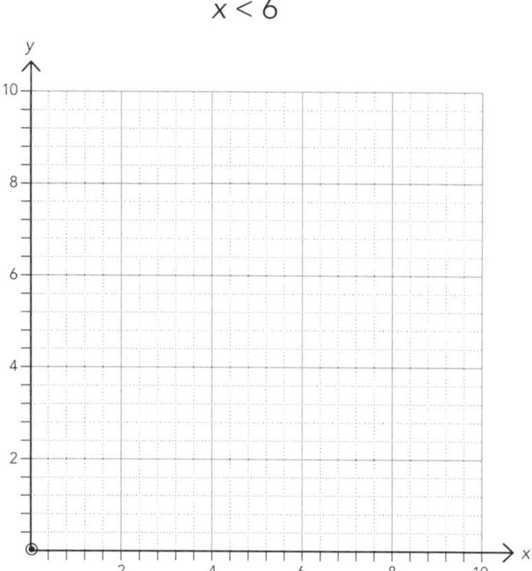

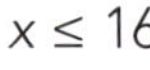

c $x \leq 16$

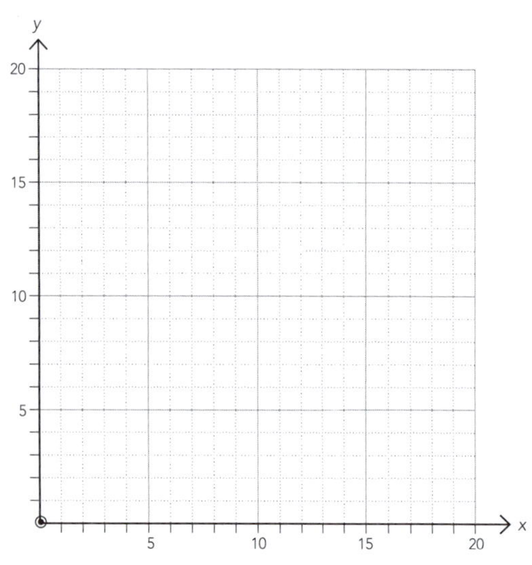

d $y > 28$

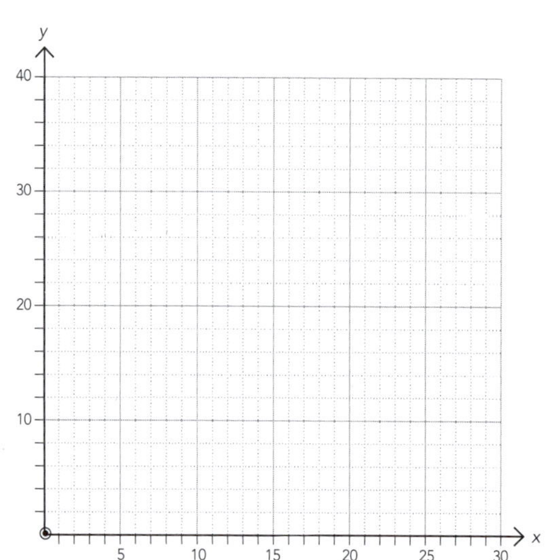

e $y \geq 45$

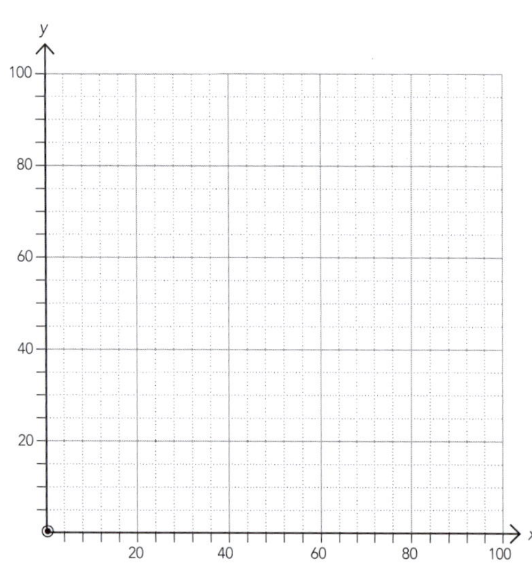

f $x > 34$

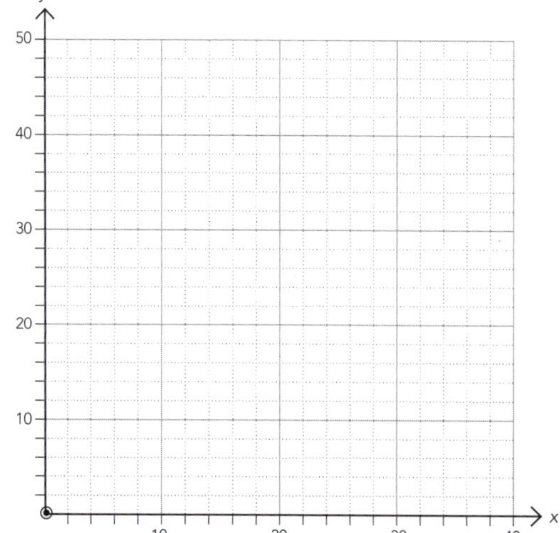

 ISBN: 9780170389396

4 Match each of the following inequations with its correct graph.

a $y \geq 0.5x + 5$ ________

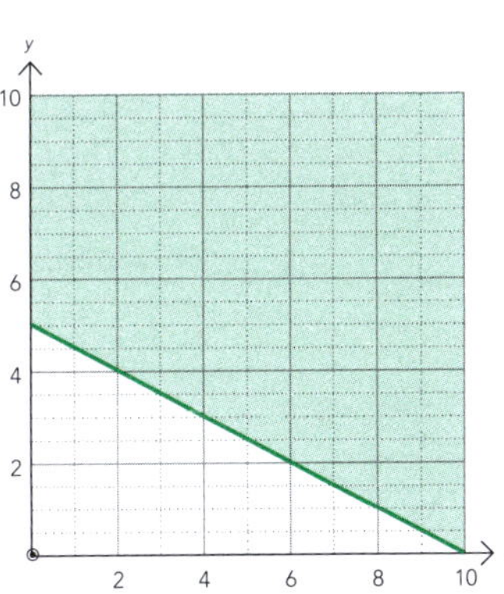

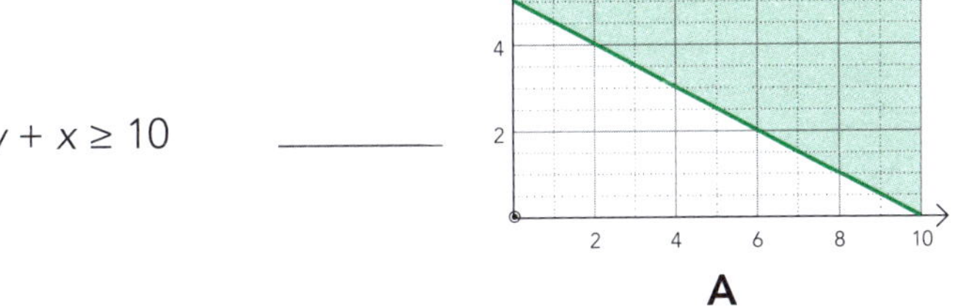

b $2y + x \geq 10$ ________

A

B

c $2y - x < 10$ ________

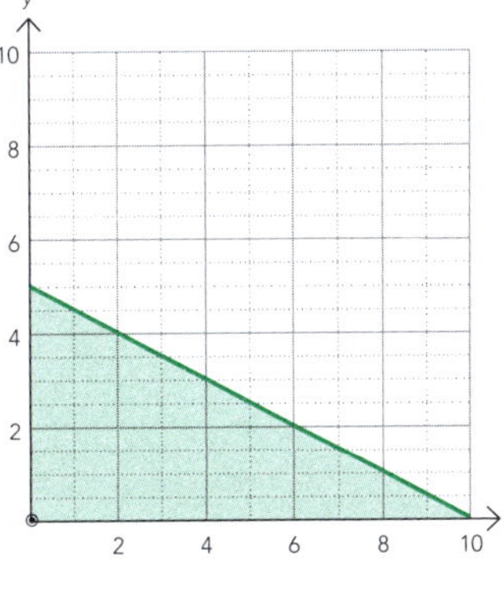

d $y \leq -0.5x + 5$ ________

C

D

e $2y + x - 10 > 0$ ________

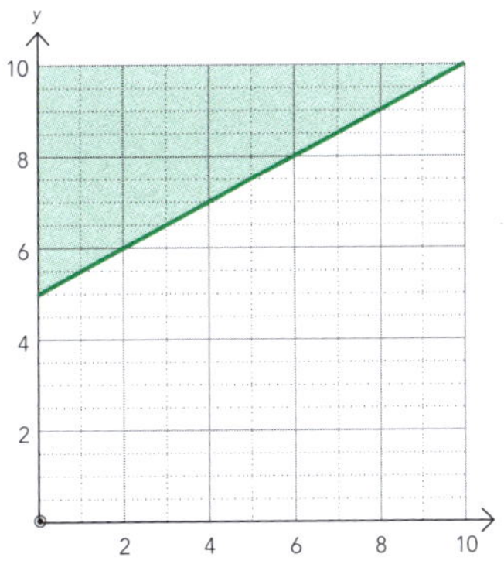

f $2y > x + 10$ ________

E

F

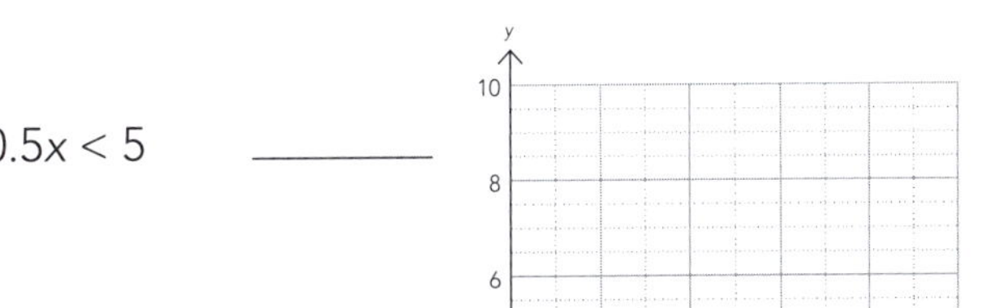

g $y + 0.5x < 5$ ________

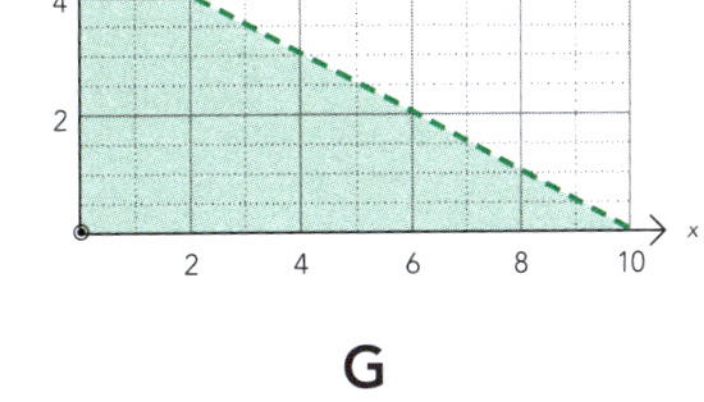

h $y - 0.5x - 5 \leq 0$ ________

G

H

ISBN: 9780170389396

5 Shade the area that is not true for the following inequations.

a $y > -x + 7$

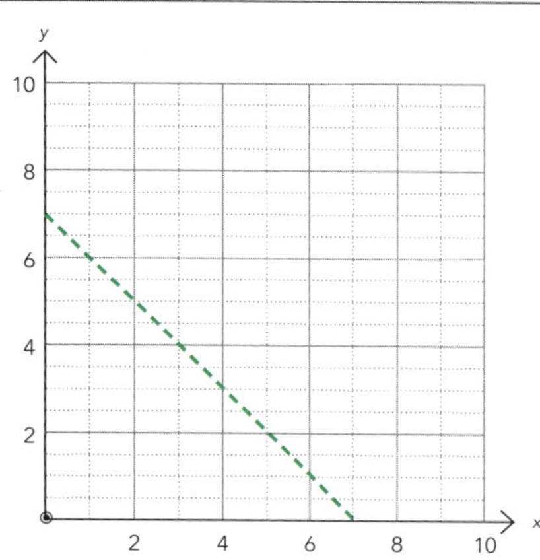

b $y \leq x$

c $2x + y \leq 12$

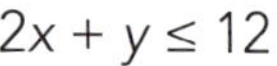

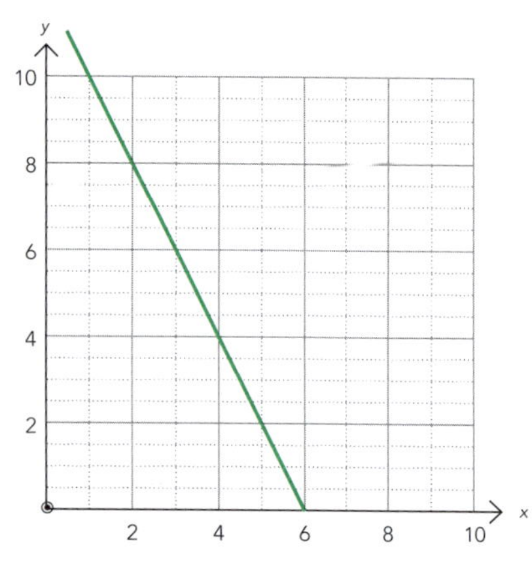

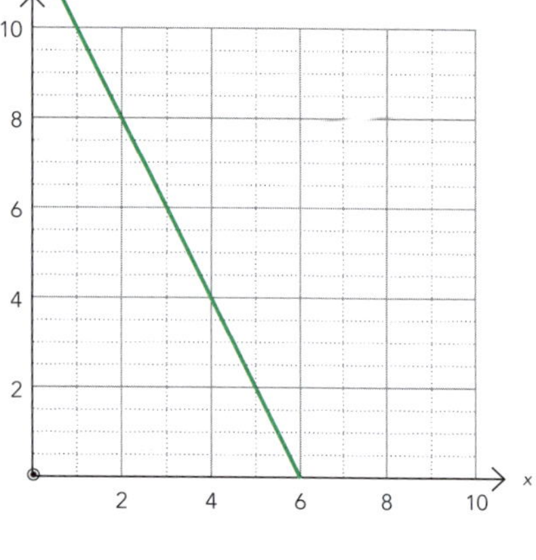

d $3x < y$

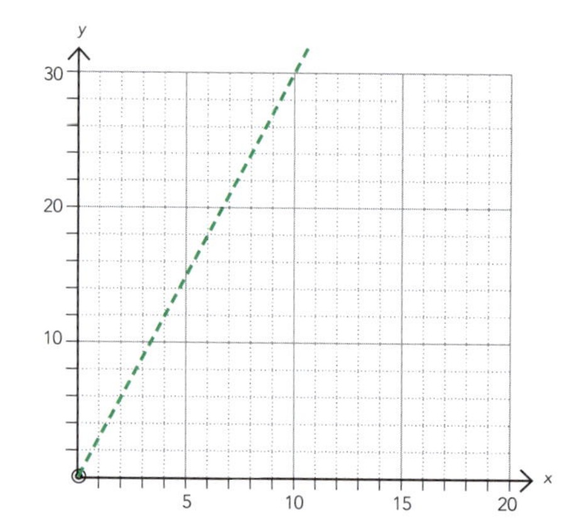

e $2y - x < 44$

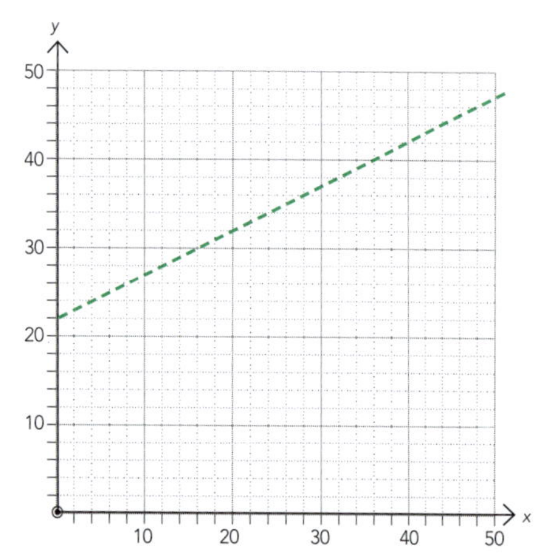

f $3y \geq 2x + 64$

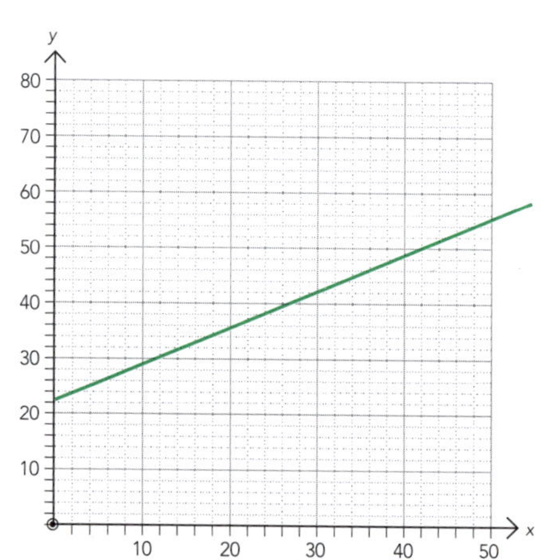

 ISBN: 9780170389396

6 Draw graphs for the following inequations, and shade the area that is not true.

a $y \geq 2x + 1$

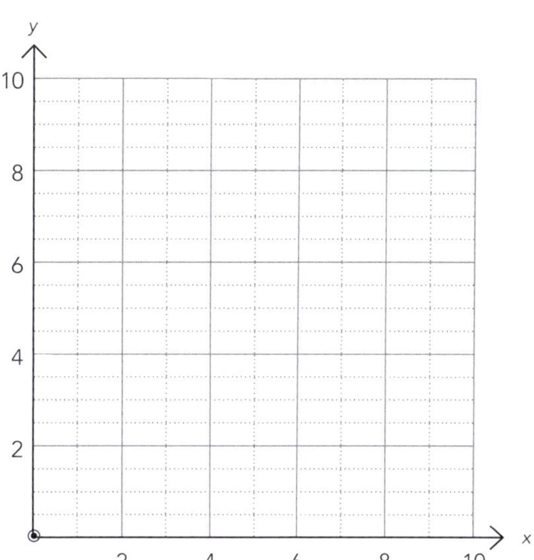

b $y < \frac{1}{2}x$

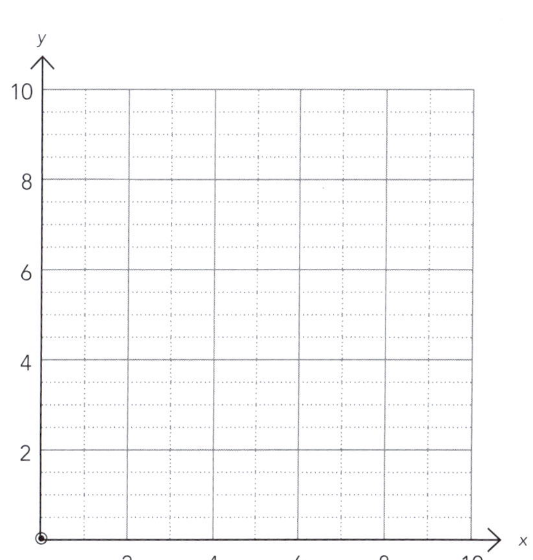

c $8x + 5y > 120$

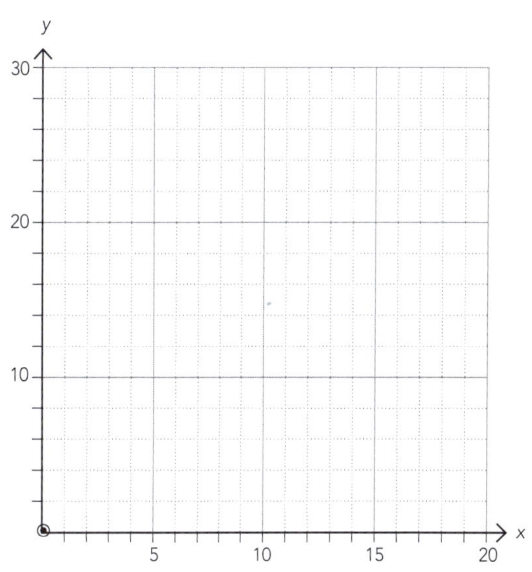

d $100x + 72y \leq 3600$

e $y - \frac{1}{2}x > 24$

f $3x \leq 288 - 4y$

 ISBN: 9780170389396

g $7x + 6y > 294$

h $84 + x \geq 3y$

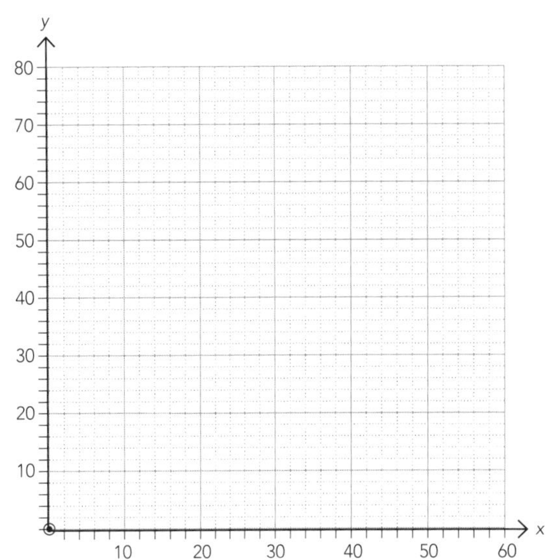

i $90 < 3b - 5a$

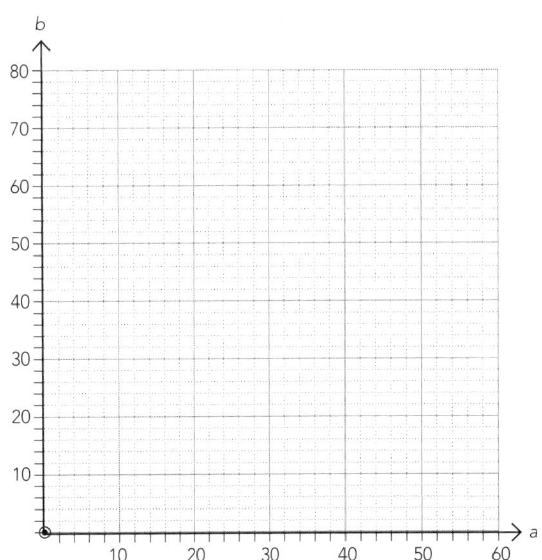

ISBN: 9780170389396

Drawing inequations using your graphics calculator

Your calculator shades the area that **is** true, rather than the area that is not true. Not all calculators will graph equations such as $x <$……, $x \geq$ ……, $x >$ ……, etc.

It is often difficult to read the values of the intercepts, but you can use the G-Solve function to find them.

Example 1: $2y + x > 6$ ← You will need to rearrange this.

$2y > 6 - x$

$y > \frac{(6 - x)}{2}$

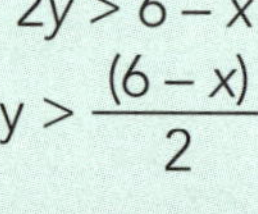

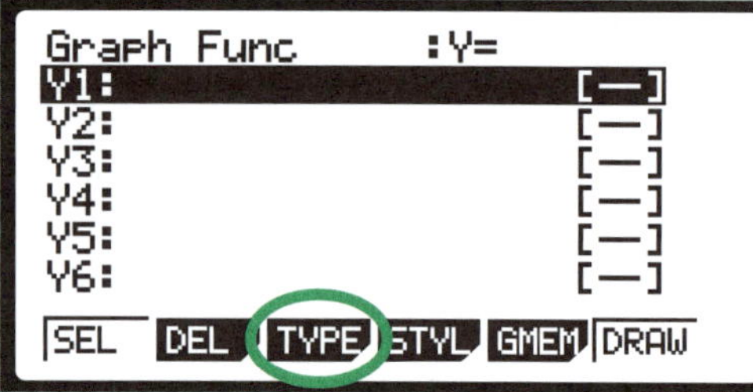

→ **Menu**
→ **Graph**
→ **F3 Type**
→ **F6 ▷**

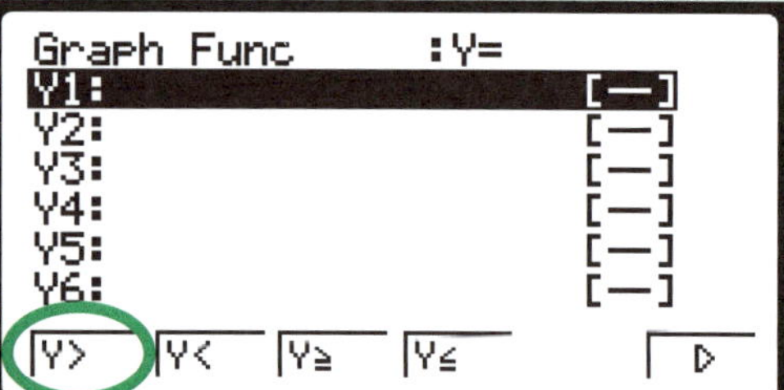

→ **F1 Y>**

Make sure this is pointing in the correct direction.

Put in the rest of the equation.

→ **Draw**

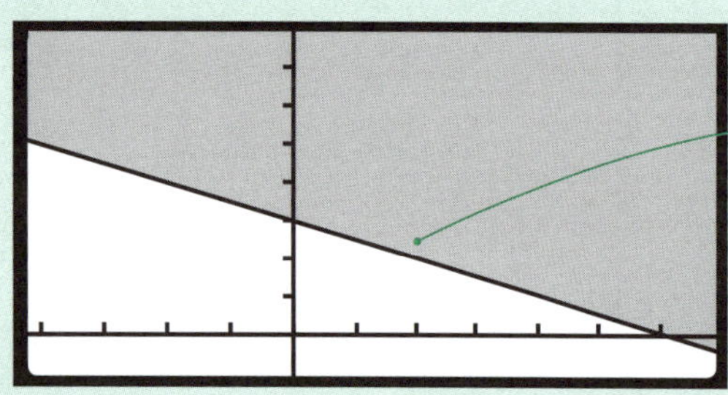

You need to be able to draw this on an axis, but remember that your calculator shades the area that is true, the **opposite** of what has been taught in this book.

Note: You will need to **draw the graph** on your own axes, and **shade the opposite** area from the one shown on your calculator.

ISBN: 9780170389396

Example 2: $x > 3$

Note: Not all models will plot $x <$ ……, $x >$ ……, etc.

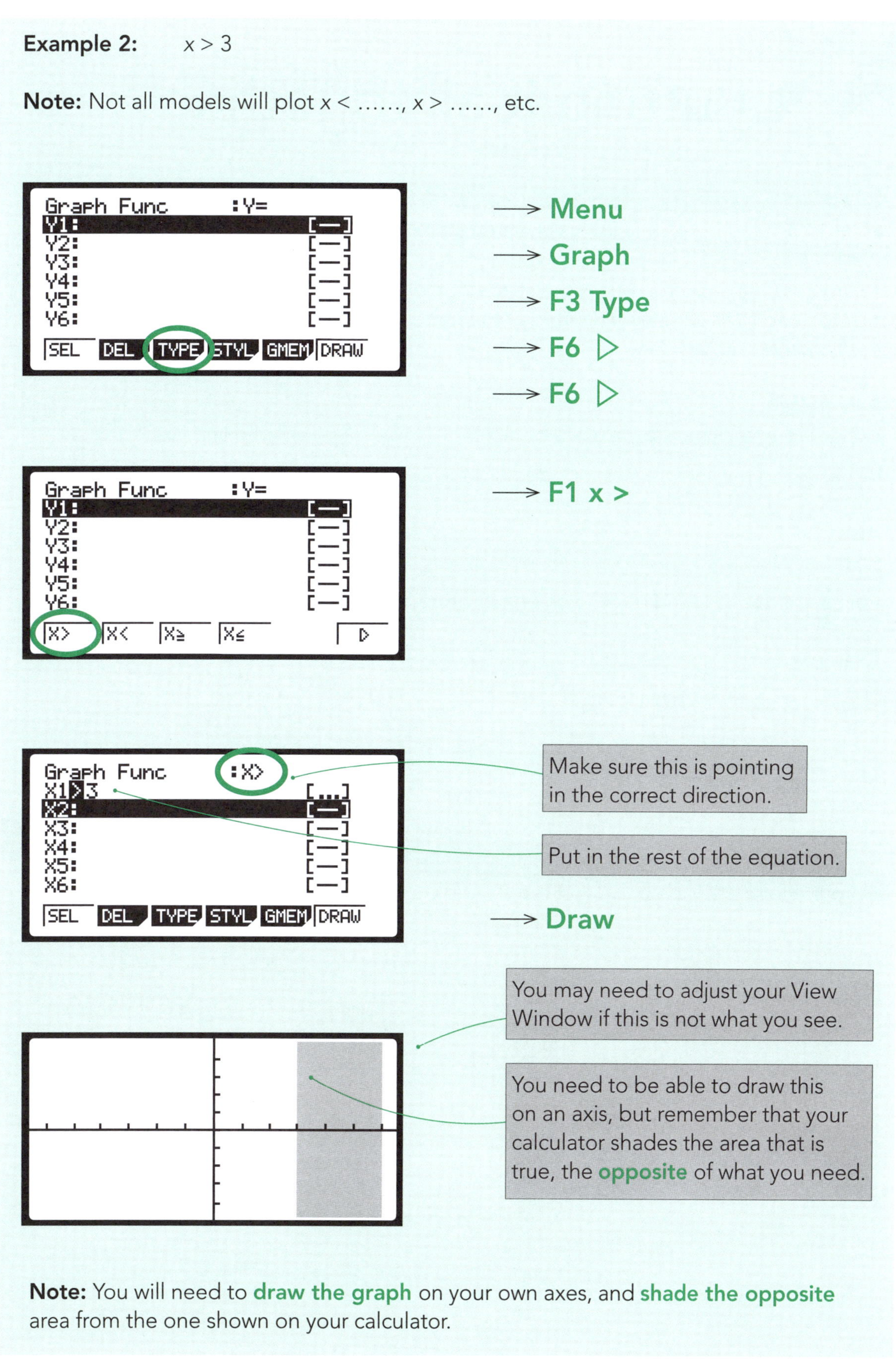

Note: You will need to **draw the graph** on your own axes, and **shade the opposite** area from the one shown on your calculator.

Putting it all together

- You will need to draw **several inequations** on one set of axes.
- You will need to identify the **area** where **all** inequations are **true**.
- This area is called the **feasible region**.
- The feasible region is usually the shape of a **polygon**.
- You need to find the **coordinates** of the **vertices** (**corners**) of the feasible region.
- You almost always deal only with **positive** values.

Remember:

Values along the line **are** included (≤ or ≥) ⇒ solid line

Values along the line **are not** included (< or >) ⇒ dashed line

Also:

Points **are** included (≤ or ≥) on all lines ⇒ solid dot

Points **are not** included (< or >) on one or more lines ⇒ hollow dot

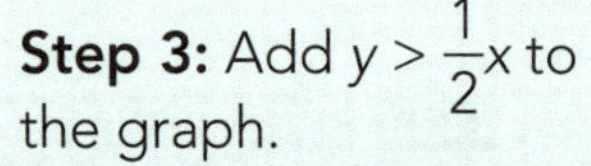

Example 1:

$x \geq 2$

$0 \leq y \leq 6$

$y > \frac{1}{2}x$

$14y + 9x < 112$

Step 1: Plot $x \geq 2$.

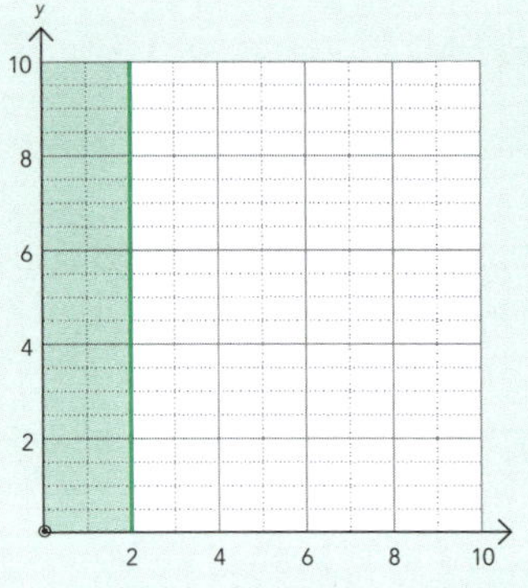

Step 2: Add $0 \leq y \leq 6$ to the graph.

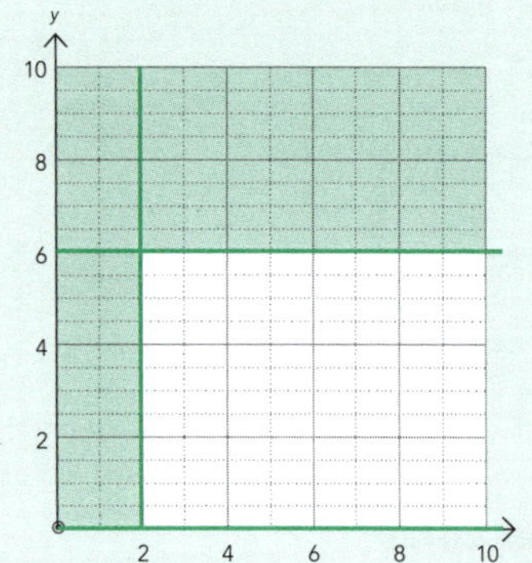

Step 3: Add $y > \frac{1}{2}x$ to the graph.

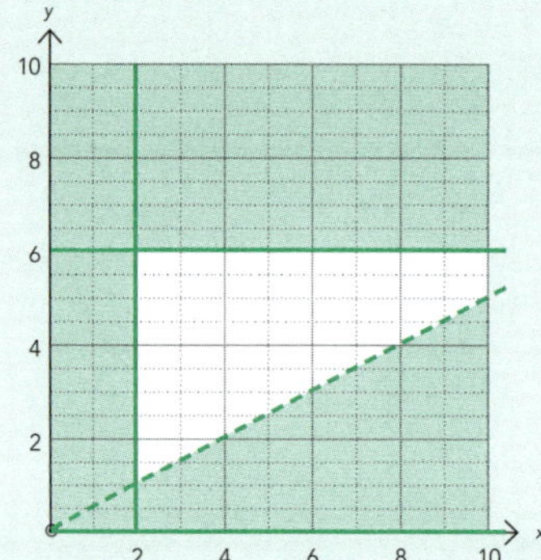

 ISBN: 9780170389396

Step 4: Add $14y + 9x < 112$ to the graph.

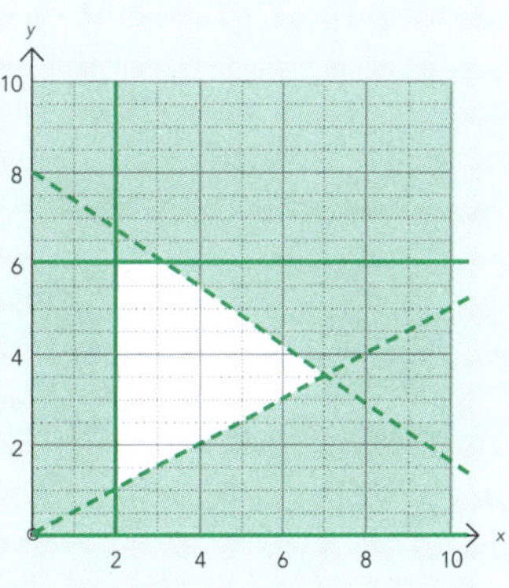

Step 5: Label the area that is not shaded.

This is the **feasible region**.

Label the **vertices** of the feasible region with **A**, **B**, **C**, etc.

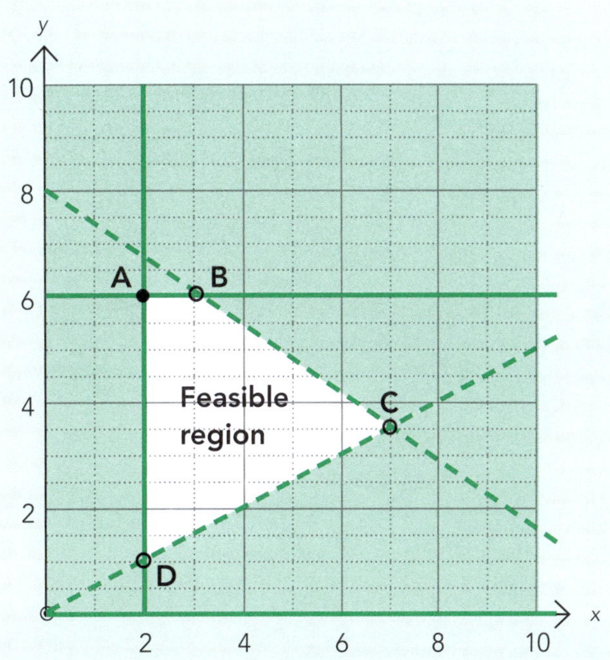

Step 6: Find the coordinates of each vertex (corner) of the feasible region. For easy ones you can read from your graph. Others will require that you use your graphics calculator or solve the relevant simultaneous equations.

A: $x = 2$ and $y = 6$ **(2, 6)**

B: $y = 6$ and $14y + 9x = 112$

$$14(6) + 9x = 112$$
$$9x = 28$$
$$x = 3.\dot{1}$$

$(3.\dot{1}, 6)$

C: $y = \frac{1}{2}x$ and $14y + 9x = 112$

$$14(\tfrac{1}{2}x) + 9x = 112$$
$$16x = 112$$
$$x = 7 \Rightarrow y = 3\tfrac{1}{2}$$

$(7, 3\frac{1}{2})$

D: $x = 2$ and $y = \frac{1}{2}x$ **(2, 1)**

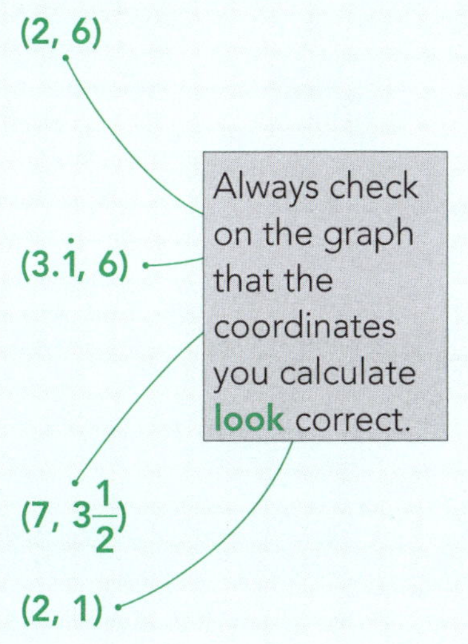

ISBN: 9780170389396

Example 2: $x \geq 0$

$y \geq 0$

$x + y > 20$

$3y + x > 48$

$6y + x > 60$

Step 1: Plot $x \geq 0$.

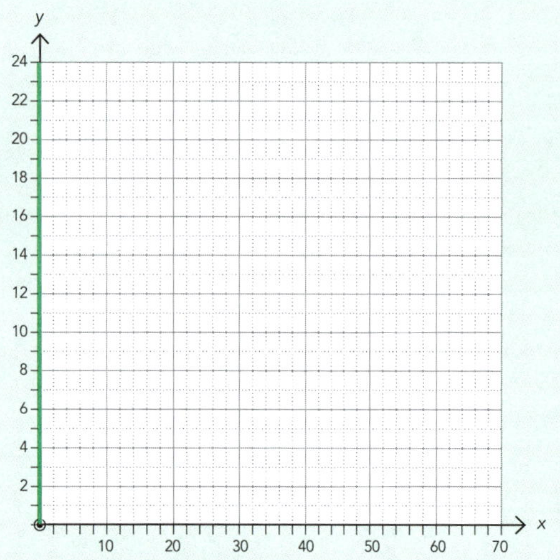

Step 2: Add $y \geq 0$ to the graph.

Step 3: Add $x + y > 20$ to the graph.

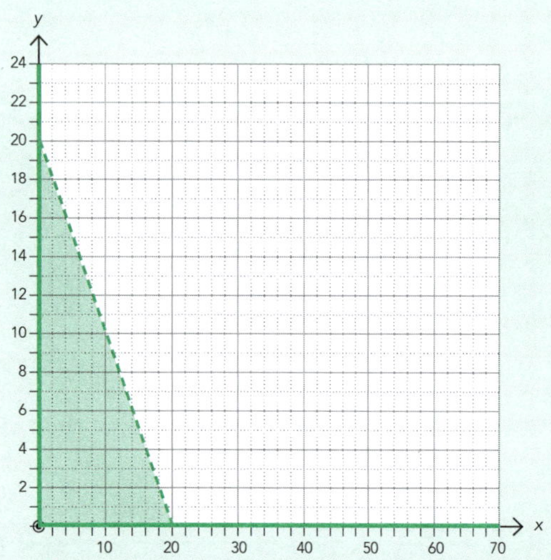

Step 4: Add $3y + x > 48$ to the graph.

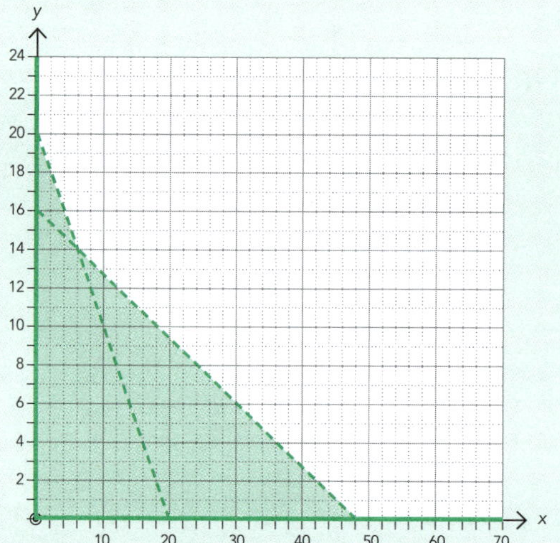

ISBN: 9780170389396

Step 5: Add $6y + x > 60$ to the graph.

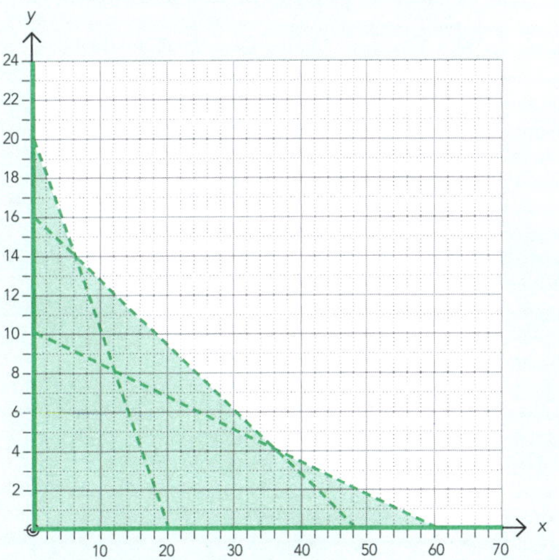

Step 6: Label the area that is not shaded.

This is the **feasible region**.

Label the **corners** of the feasible region with **A**, **B**, **C**, etc.

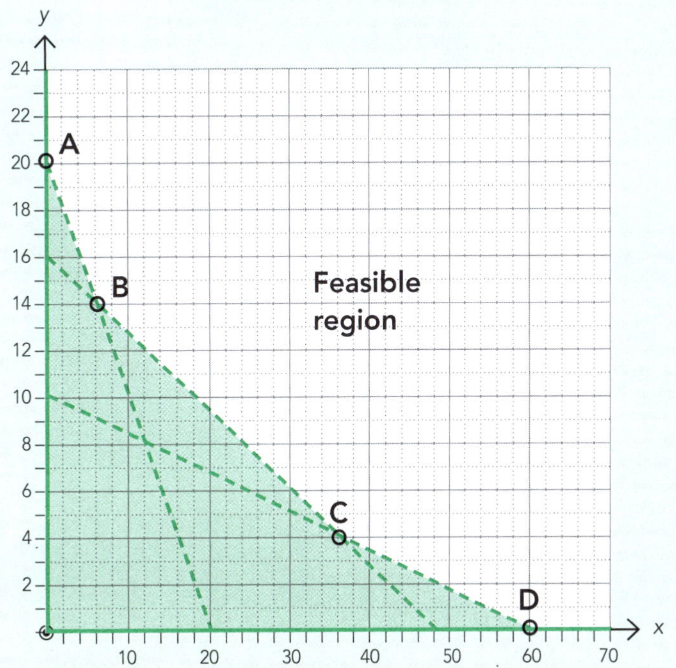

Step 7: Find the coordinates of each corner of the feasible region.
For easy ones you can read from your graph. Others will require that you use your graphics calculator or solve the relevant simultaneous equations.

A: $x = 0$ and $y = 20$ **(0, 20)**

B: $x + y = 20$ and $3y + x = 48$

$$3(20 - x) + x = 48$$
$$60 - 2x = 48$$
$$x = 6 \Rightarrow y = 14$$

(6, 14)

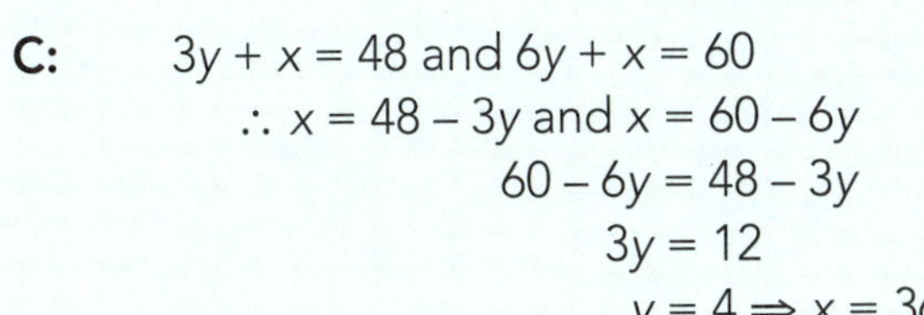

C: $3y + x = 48$ and $6y + x = 60$

$$\therefore\ x = 48 - 3y \text{ and } x = 60 - 6y$$
$$60 - 6y = 48 - 3y$$
$$3y = 12$$
$$y = 4 \Rightarrow x = 36$$

(36, 4)

D: $6y + x = 60$ and $y = 0$ **(60, 0)**

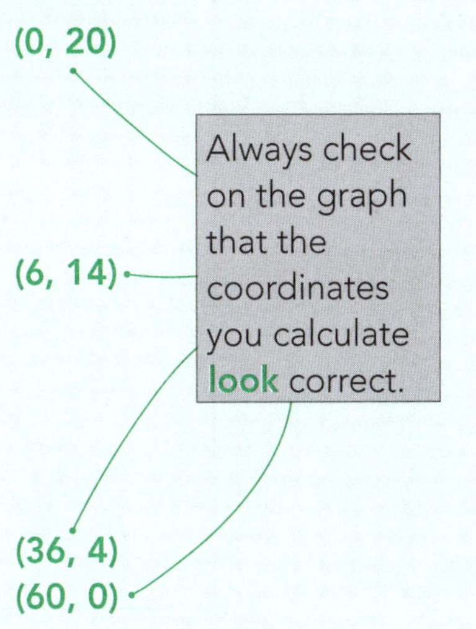

ISBN: 9780170389396

Draw the following sets of inequations on one set of axes, then find and label the feasible region and the coordinates of its corners.

1

$$x \geq 1$$
$$0 \leq y \leq 9$$
$$y > x$$
$$2y + 3x < 30$$

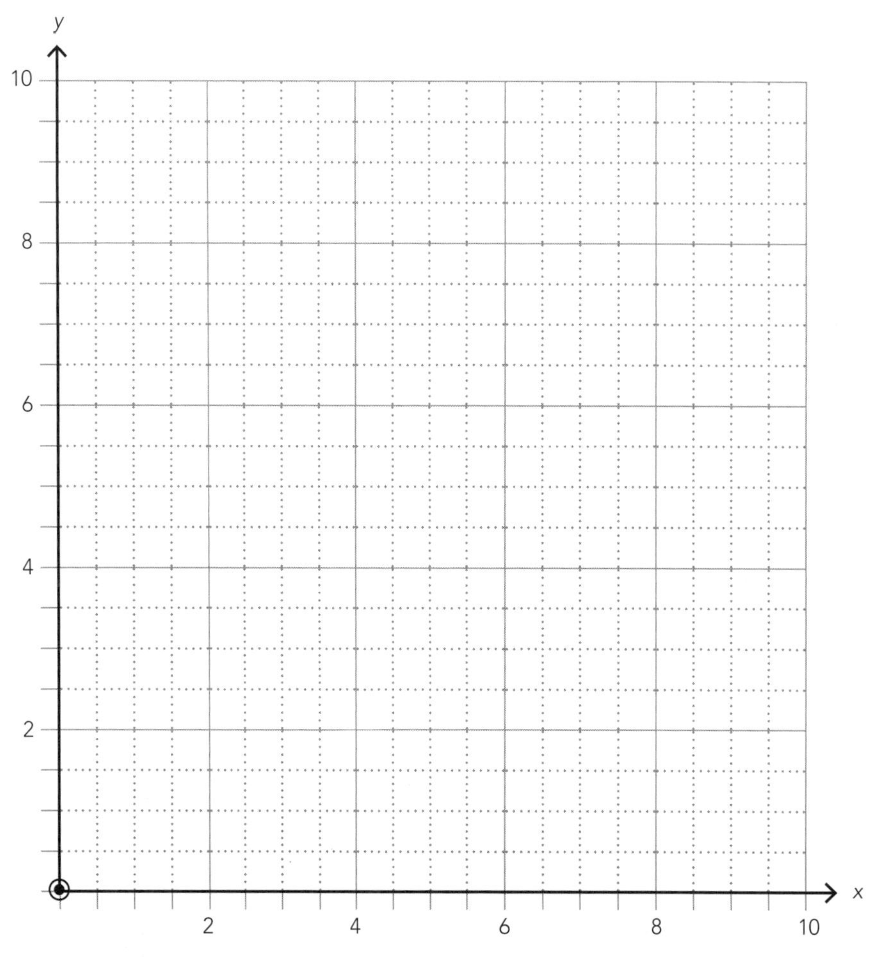

ISBN: 9780170389396

2

$$x \geq 0$$
$$y \geq 0$$
$$y + 2x > 10$$
$$2y + x > 14$$
$$y - x + 2 > 0$$

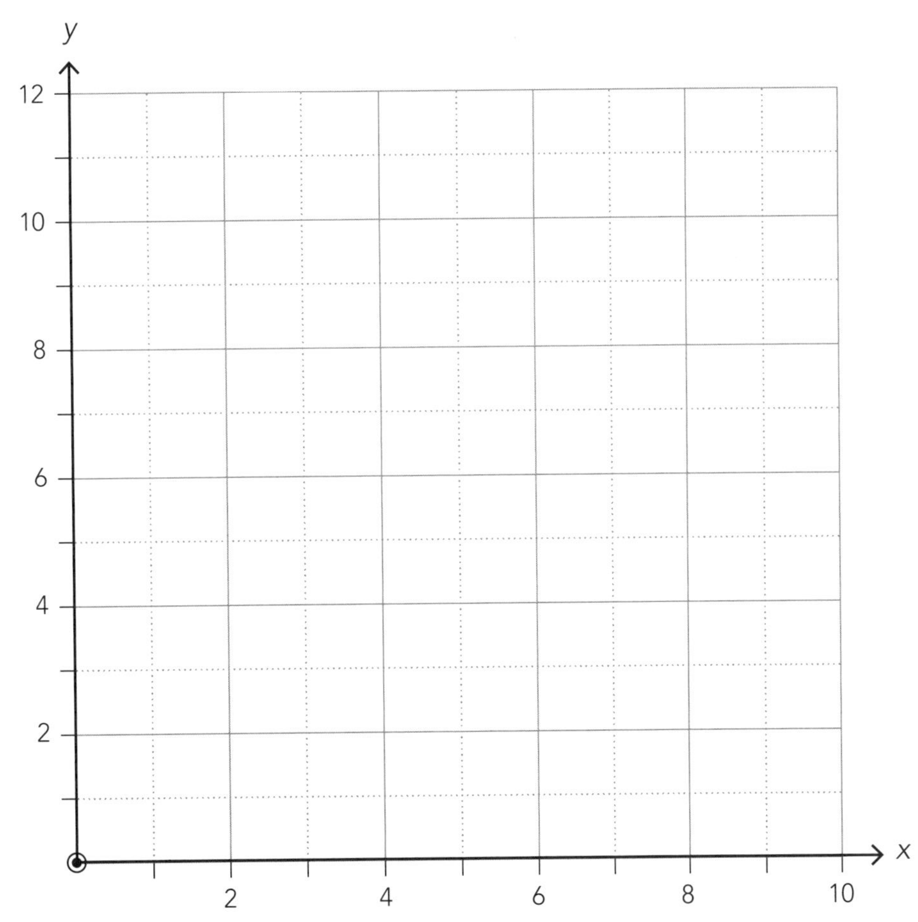

ISBN: 9780170389396

3

$$0 \le x < 30$$
$$0 \le y \le 36$$
$$4y + x > 40$$
$$y + x - 22 > 0$$
$$y > -3x + 38$$

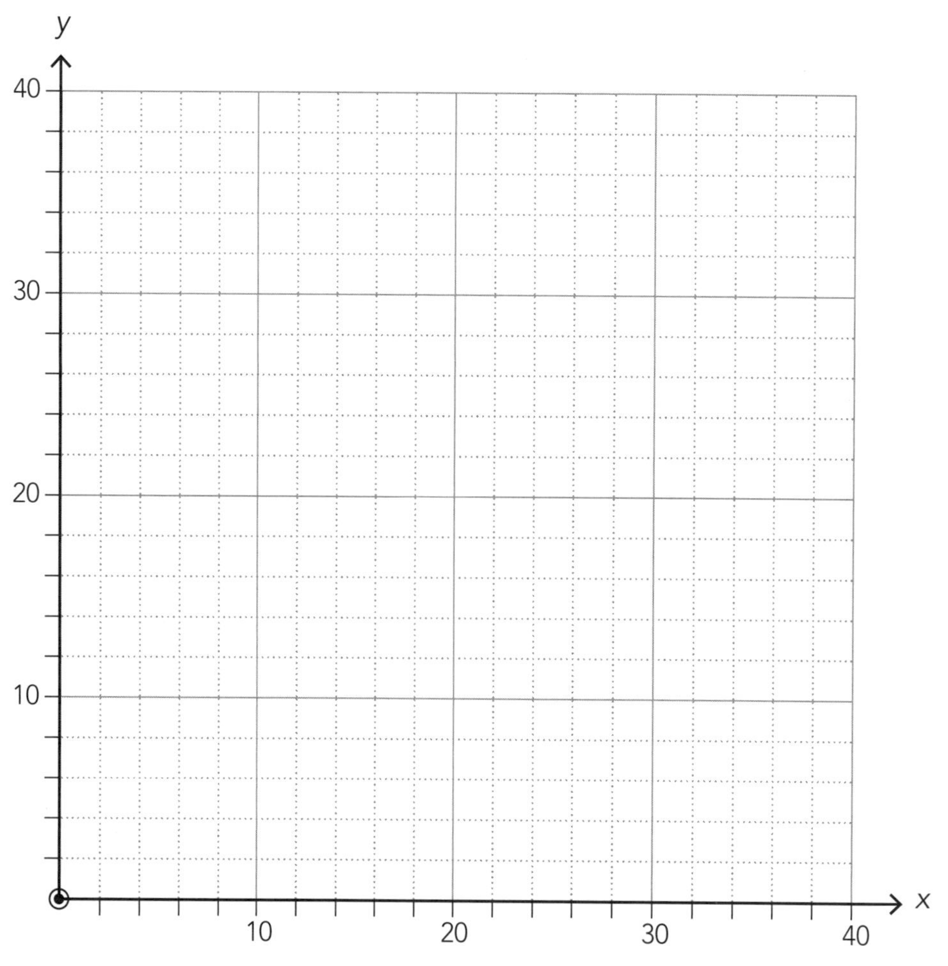

ISBN: 9780170389396

4

$$x \geq 8$$
$$y \geq 8$$
$$6y - x < 120$$
$$3y + 2x < 120$$
$$y < -2x + 88$$

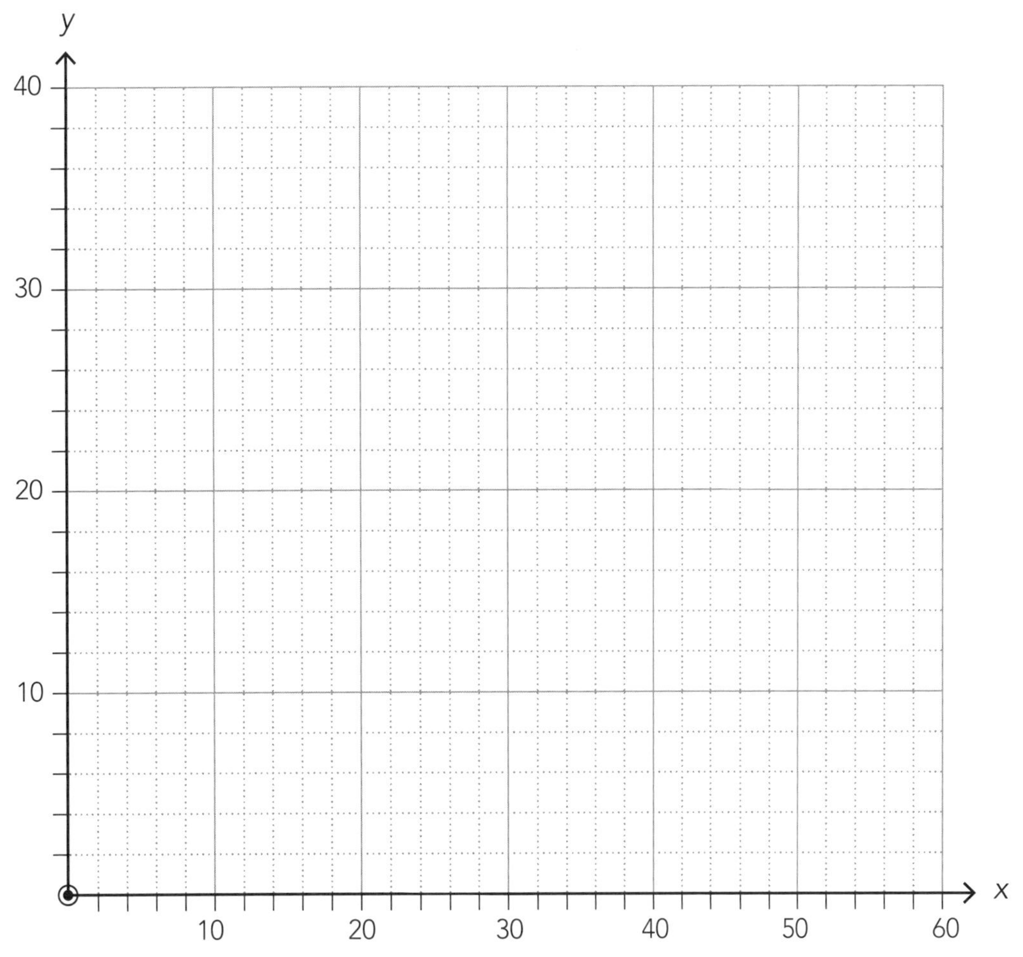

ISBN: 9780170389396

5

$$x \geq 0$$
$$y \geq 0$$
$$y > -4x + 96$$
$$5y + x > 100$$
$$5y + 6x \geq 340$$
$$2y + x > 94$$

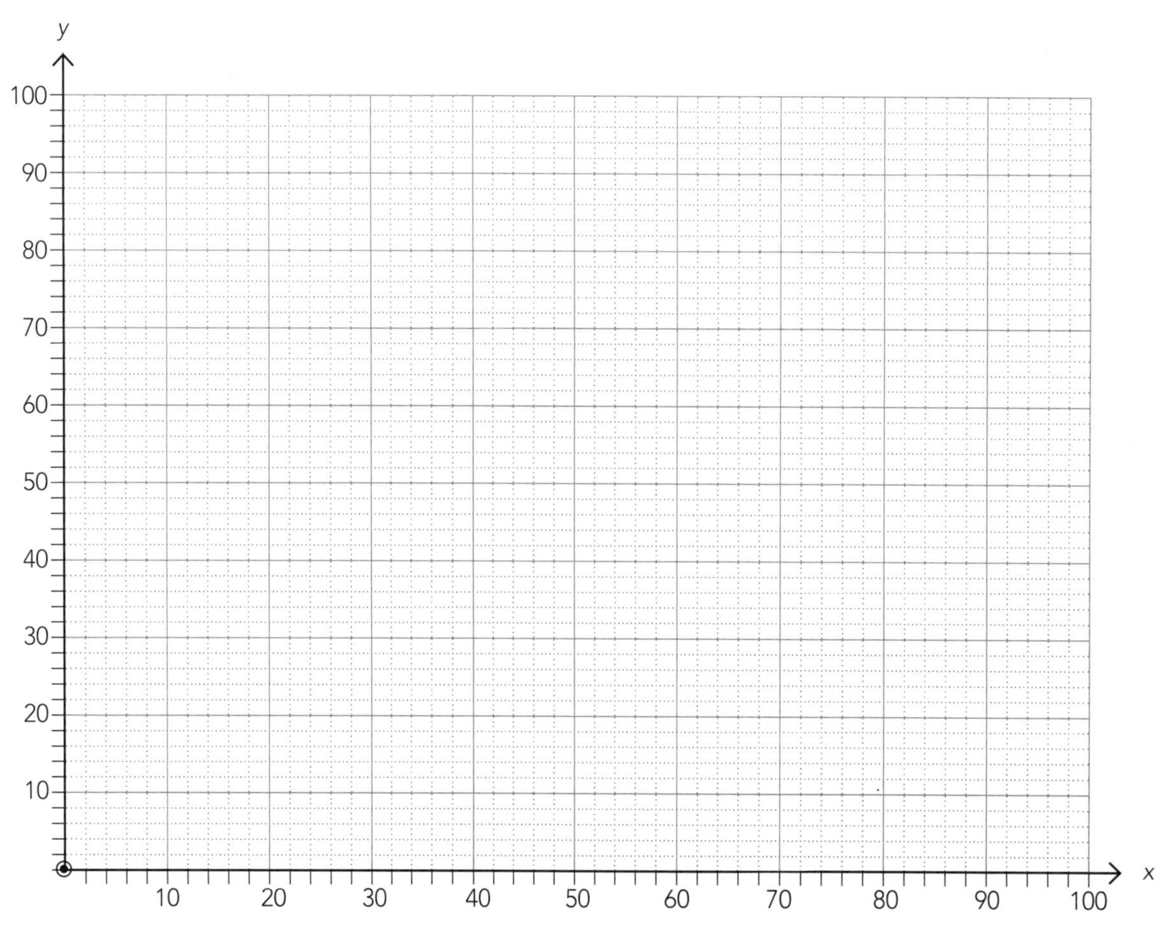

 ISBN: 9780170389396

6

$$0 \le x \le 98$$
$$0 \le y \le 92$$
$$y + 2x > 100$$
$$x + y > 84$$
$$2y + x \ge 90$$

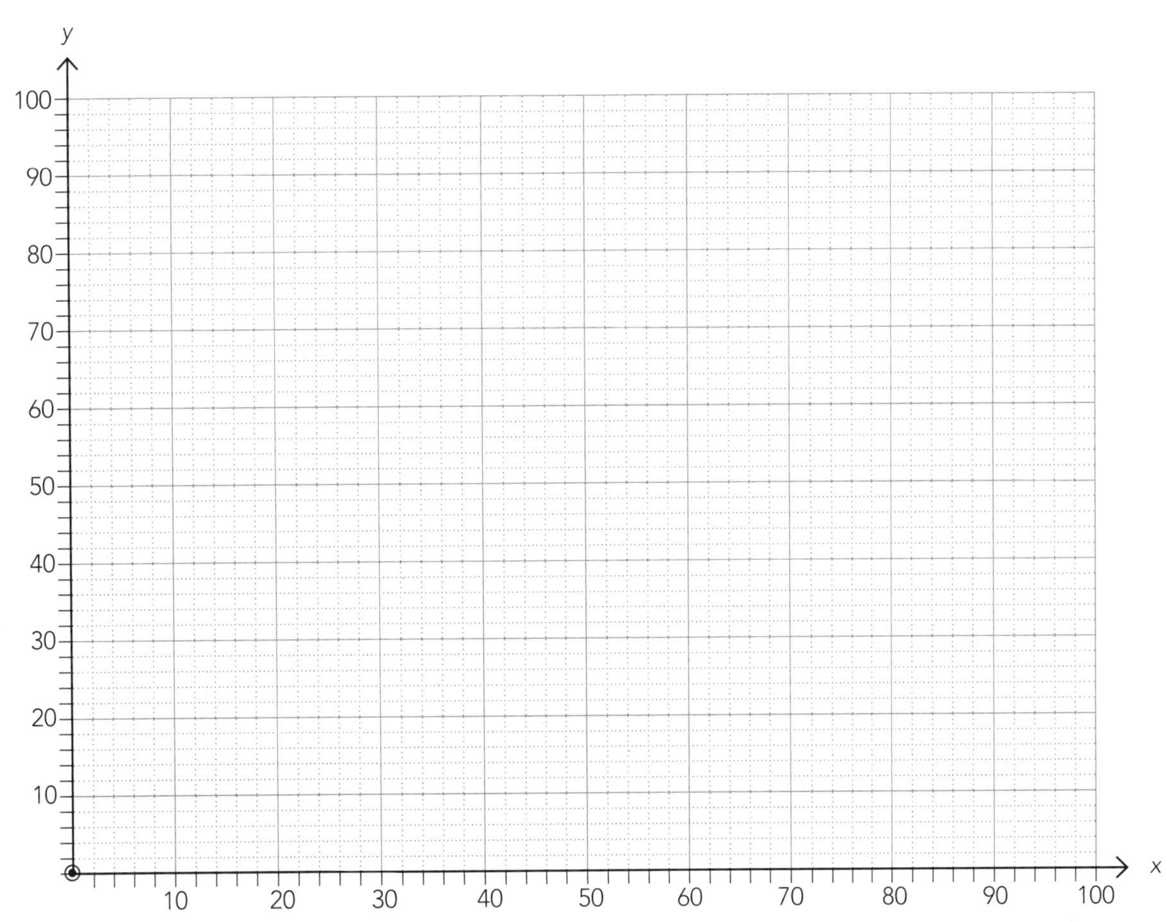

ISBN: 9780170389396

Writing inequations

- Always **define your variables** before writing an inequation.
- If you are not sure whether your inequation is correct, **test** it by substituting some numbers.

Examples:

A canteen sells stuffed potatoes and cups of soup.

Let x = the **number** of stuffed potatoes sold
y = the **number** of cups of soup sold

> Make sure you state **what it is** about the stuffed potatoes and cups of soup that you are talking about. Do **not** just write 'x = stuffed potatoes'.

a The total of number of cups of soup and stuffed potatoes sold is at least 90.

$x + y \geq 90$

> **Test:** If it sold 50 of each, they would total at least 90.
> $50 + 50 \geq 90$
> $\therefore 100 \geq 90$
> The last line is true, so the inequation works.

b It sells stuffed potatoes for $3 and cups of soup for $2. The total value of the cups of soup and stuffed potatoes on Monday was more than $120.

$3x + 2y > \$120$

> **Test:** If it sold 20 stuffed potatoes and 31 cups of soup, that would add to more than $120.
> $3 \times 20 + 2 \times 31 > 120$
> $\therefore 60 + 62 > 120$
> The last line is true, so the inequation works.

c For every stuffed potato sold, it sells at least two cups of soup.

> Be **VERY** careful of this type!

$y \geq 2x$

> **Test:** If it sold 50 stuffed potatoes, it would sell at least 100 cups of soup.
> $100 \geq 2 \times 50$
> $\therefore 100 \geq 100$
> The last line is true, so the inequation works.

d It sells at least 23 more cups of soup than stuffed potatoes.

$y \geq x + 23$

> **Test:** If it sold 50 stuffed potatoes, it would sell 73 or more cups of soup.
> $73 \geq 50 + 23$
> $\therefore 73 \geq 73$
> The last line is true, so the inequation works.

ISBN: 9780170389396

Write an inequation for each of the following situations.

1 The canteen also sells packets of sandwiches and juicies. Let x represent the number of packets of sandwiches and y represent the number of juicies sold.

Remember – **you** will need to do this!

a It sells a maximum of 15 packets of sandwiches.

b It sells more than 40 juicies.

c The total of number of packets of sandwiches and juicies sold is no more than 70.

d Juicies cost \$2 and packets of sandwiches cost \$4. The total value of juicies and packets of sandwiches is \$150 at most.

e It sells at least three times as many juicies as packets of sandwiches.

f It sells 30 fewer packets of sandwiches than it does juicies.

g The difference between the number of juicies sold and the number of packets of sandwiches sold is more than 20.

ISBN: 9780170389396

2 A company sells motor scooters and electric bikes. Let x represent the number of scooters sold and y represent the number of bikes sold.

a The number of bikes sold exceeds the number of scooters sold.

b It sells a minimum of 150 scooters.

c It sells between 500 and 600 bikes.

d Scooters cost \$2400 and electric bikes cost \$800. The total value of bikes and scooters sold during a period was between \$50 000 and \$60 000 inclusive.

e The sum of the scooters and bikes sold is 700 or fewer.

f In one week, it sells over twice as many bikes as it sells scooters.

g In another week, the number of bikes sold is not less than three times the number of scooters sold.

ISBN: 9780170389396

Write several inequations for each of the following situations.

3 Let x represent the number of dogs and y represent the number of cats. For every dog there are fewer than three cats. There is a total of at least 20 cats and dogs, but fewer than 30.

4 A farmer has a maximum of 1600 pigs and sheep altogether. There are at least five times as many sheep as pigs. The difference between the number of sheep and pigs is not less than 900.
Let x represent the number of pigs and y represent the number of sheep.

5 An appliance dealer sells more than double the number of toasters as kettles. Kettles sell for $85 and toasters sell for $75. The total value of kettles and toasters sold is no more than $1500, but more than $1000. The total value of the toasters sold is at least $400 more than the total value of the kettles sold.
Let x represent the number of toasters sold and y represent the number of kettles sold.

ISBN: 9780170389396

Linear programming

- When tackling a linear programming problem you must identify the **constraints** and the **objective** function.
- The **constraints** are the statements that need to be converted into **inequalities**.
- The **objective function** is an expression for whatever has to be **maximised** or **minimised**, usually profit.

Example:

A new branch of a car hire company is being set up. They are buying conventional cars (run by petrol or diesel) and electric cars.

- They are allowed a maximum of 100 cars altogether.
- Due to demand, they do not want more than 46 electric cars and they do not want more than 60 conventional cars.
- The head office insists that no more than 50% of the branch's fleet is electric.
- They also insist that for every electric car, there should be at most four conventional cars.
- They will make an annual profit of $3000 per conventional car and $3500 per electric car.

How many of each should they buy in order to maximise their profit?

Step 1: Identify the variables.

Look for the question. In this case 'How many of each should they buy in order to maximise their profit?'

Let x represent the number of conventional cars and y represent the number of electric cars.

This tells you that the variables will be numbers of conventional cars and electric cars.

Step 2: Find the information needed to write the objective function.

Hints:
1. This often contains the words maximum or minimum.
2. It is usually in the form of a question.
3. It is usually at the end of the question.
4. It often involves profit.
5. It will usually take the form $P = ax + by$.

In this case: 'They will make an annual profit of $3000 per conventional car and $3500 per electric car. How many of each should they buy in order to maximise their profit?'

$$Profit = 3000x + 3500y$$

Step 3: Write the constraints.

'They are allowed a maximum of 100 cars altogether.'	$x + y \leq 100$
'Due to demand, they do not want more than 46 electric cars …'	$0 \leq y \leq 46$
'… and they do not want more than 60 conventional cars.'	$0 \leq x \leq 60$
'The head office insists that no more than 50% of the branch's fleet is electric.'	$y \leq x$
'They also insist that for every electric car, there should be at most four conventional cars.'	$y \geq \frac{1}{4}x$

Test some numbers if you are not sure about this equation:
10 electric ⇒ 40, 39, 38, etc. conventional cars.

 ISBN: 9780170389396

Step 4: Graph the constraints and find the feasible region.

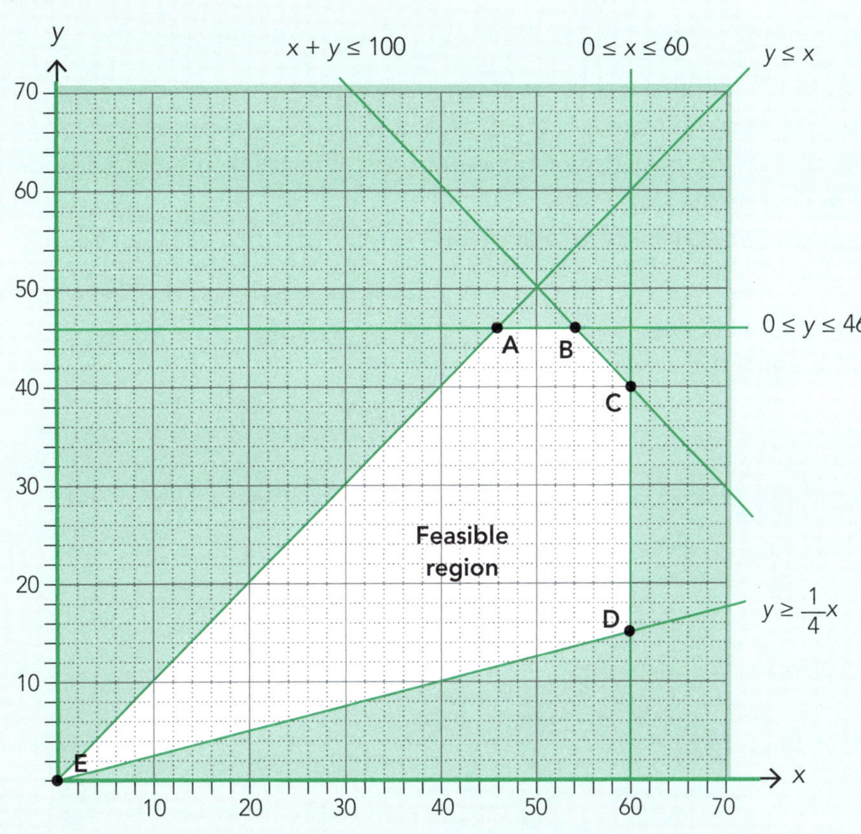

Step 5: List the coordinates of corners of the feasible region.

A: (46, 46)
B: (54, 46)
C: (60, 40)
D: (60, 15)
E: (0, 0)

Step 6: Evaluate the objective function (*Profit* = 3000*x* + 3500*y*) at each corner:

A: (46, 46): $P = 3000(46) + 3500(46) = \$299\,000$
B: (54, 46): $P = 3000(54) + 3500(46) = \$323\,000$
C: (60, 40): $P = 3000(60) + 3500(40) = \$320\,000$
D: (60, 15): $P = 3000(60) + 3500(12) = \$232\,500$
E: (0, 0): $P = 3000(0) + 3500(0) = \$0$

Step 7: State which point is optimal (either the maximum or minimum).

∴ The largest profit ($323 000) is made at point B (54, 46).

Step 8: Write a sentence describing this result in context.

The maximum profit the branch can make is **$323 000**, which they get by buying **54** conventional cars and **46** electric cars.

Make sure you include all the relevant **numbers** in your sentence.

ISBN: 9780170389396

Answer the following questions.

1 The members of the school council are making prize packs for the talent quest.

- They need to make a total of at least 24 prizes.
- There must be at least 6 first prizes and 12 runner-up prizes.
- The number of first prizes must be at most, five more than the number of runner-up prizes.
- Each first prize will cost $25 and each runner-up prize will cost $8.

How many of each should they make in order to minimise the cost of prizes?

Step 1: Identify the variables.

Step 2: Write the objective function.

Step 3: Write the constraints.

Step 4: Graph the constraints and label the feasible region.

Step 5: List the coordinates of corners of the feasible region.

ISBN: 9780170389396

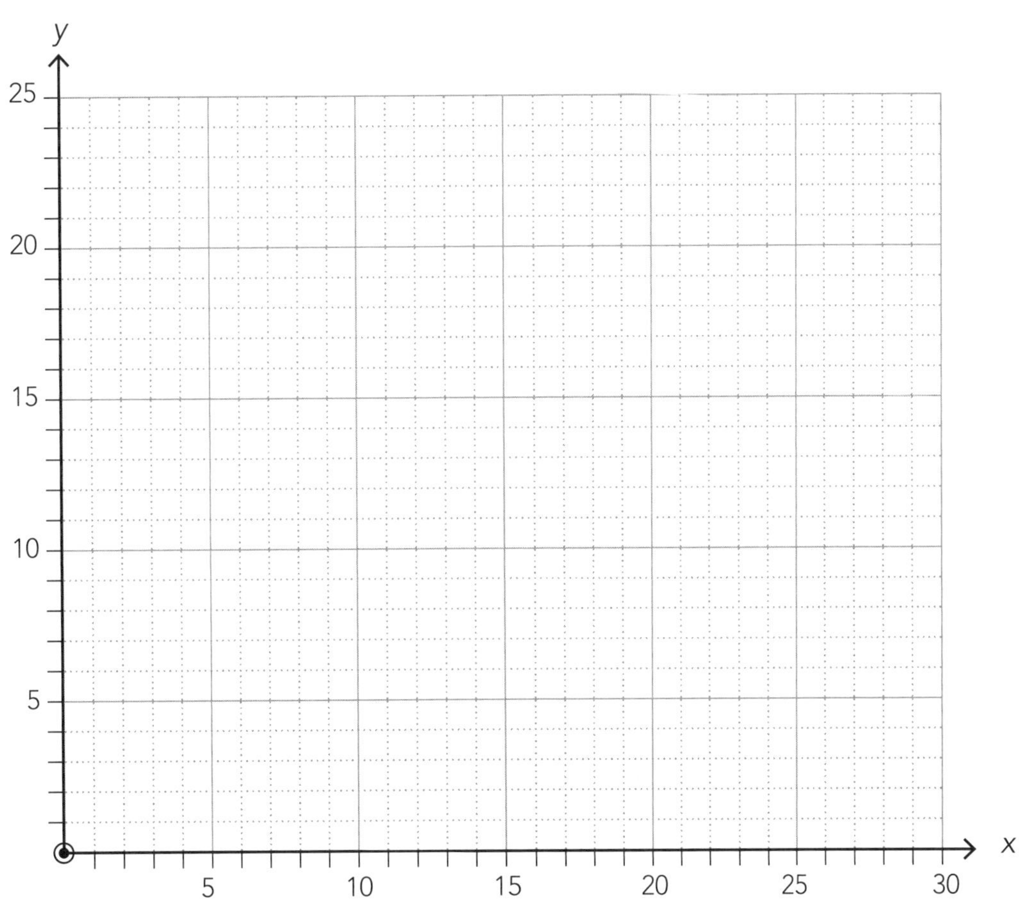

Step 6: Evaluate the objective function at each corner.

Step 7: State which point is optimal (either the maximum or minimum).

Step 8: Write a sentence describing this result in context.

2 The local council is planning a mass planting of flowers in their regional colours (yellow and red) for their summer festival. They have chosen sunflowers and poppies for the mass planting.

- In order to maintain a balance of colours, they don't want more than 150 of either flower.
- Sunflower seedlings cost 50 cents each and poppy seedlings cost 30 cents each. The cost of seedlings must not exceed $90.
- They have a maintenance budget of $480 for this area of planting. It costs $3 to maintain each sunflower plant and $1 for each poppy plant.
- The visual impact of sunflowers is twice the visual impact of poppies.

Calculate the number of poppies that the council should plant in order to maximise the visual impact of the mass planting.

Step 1: Identify the variables.

Step 2: Write the objective function.

Step 3: Write the constraints.

Step 4: Graph the constraints and label the feasible region.

Step 5: List the coordinates of corners of the feasible region.

ISBN: 9780170389396

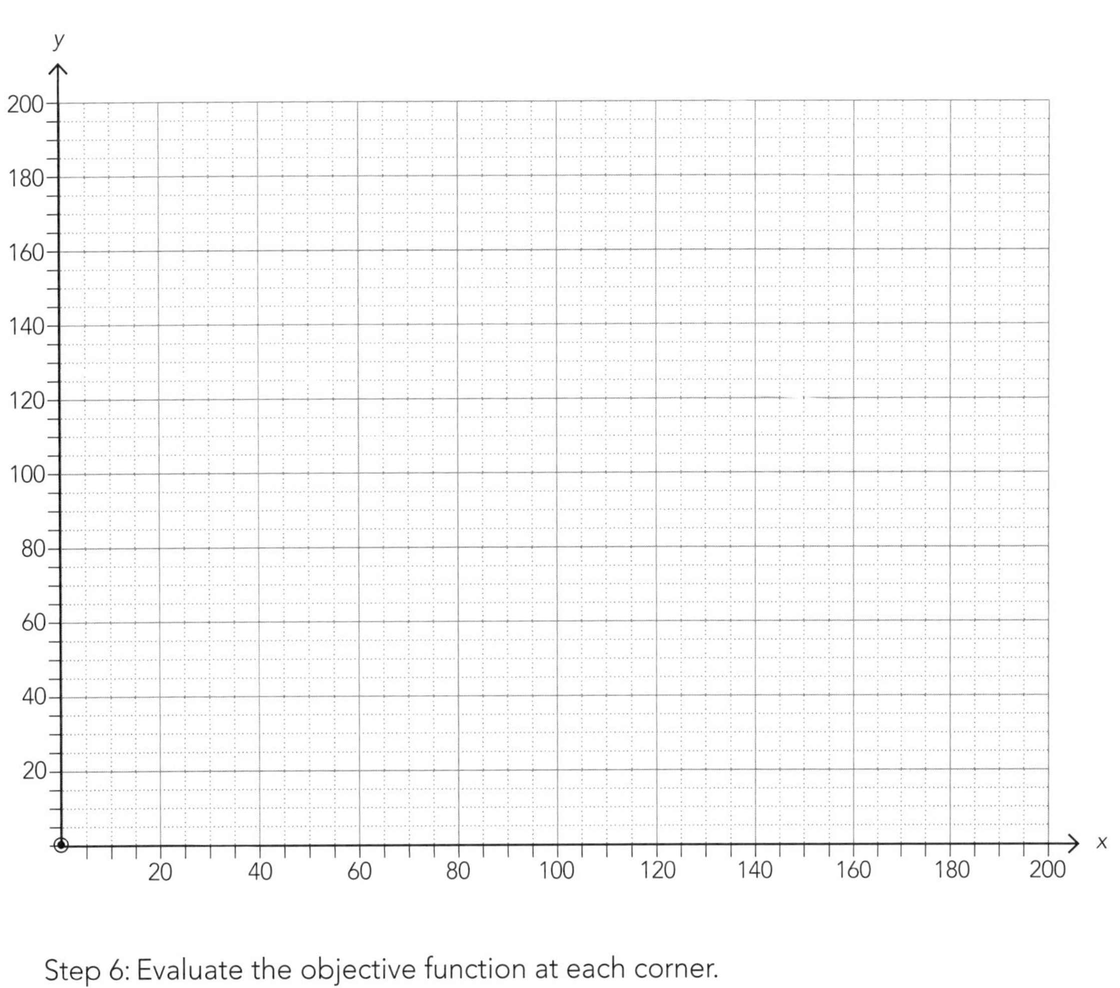

Step 6: Evaluate the objective function at each corner.

Step 7: State which point is optimal (either the maximum or minimum).

Step 8: Write a sentence describing this result in context.

ISBN: 9780170389396

3 A sports trip is being planned, and the students will be transported in parents' vans and cars.

- There are 15 adults (parents and teachers) who are prepared to drive a vehicle.
- They need to take at least 4 vans in order to fit sports equipment.
- Two parents are willing to drive their own cars, but not vans.
- Each van must contain at least 2 Year 13 students, and each car must contain at least 1 Year 13 student. The minimum number of Year 13 students going on the trip is 16.
- The maximum number of students that can travel in each van is 10, and in each car it is 4.

Calculate the maximum and minimum numbers of students that can go on the sports trip. Describe any limitation on taking the maximum number of students.

Step 1:

Step 2:

Step 3:

Step 4:

Step 5:

ISBN: 9780170389396

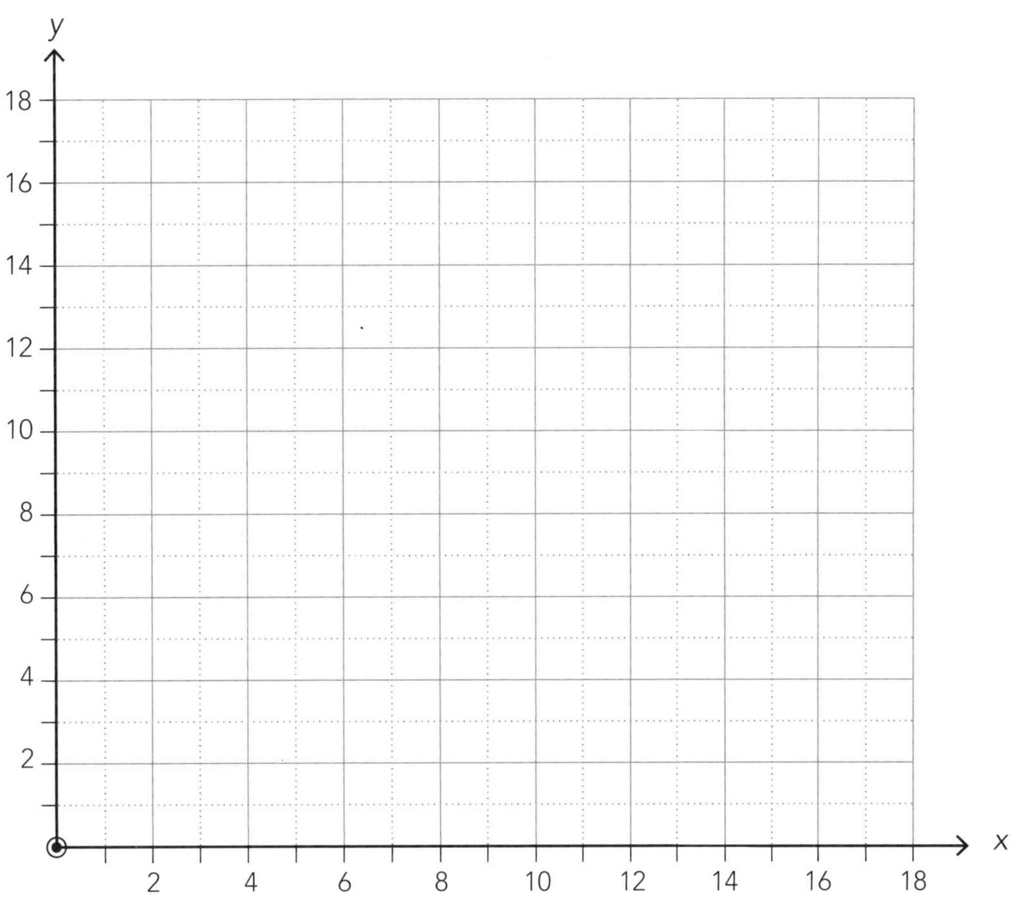

Step 6:

Step 7:

Step 8:

4 Clare is making tiny bunches of flowers to go in the boys' buttonholes for the school formal.

- She has 20 fern leaves, 57 rosebuds and 54 lavender sprigs.
- She makes two different designs.
- Design A needs 1 fern leaf, 3 rosebuds and 2 lavender sprigs.
- Design B needs 1 fern leaf, 2 rosebuds and 3 lavender sprigs.
- She charges \$5 for design A and \$3 for design B.

How many of each design should she make in order to maximise her profit?
What will she have left over?

ISBN: 9780170389396

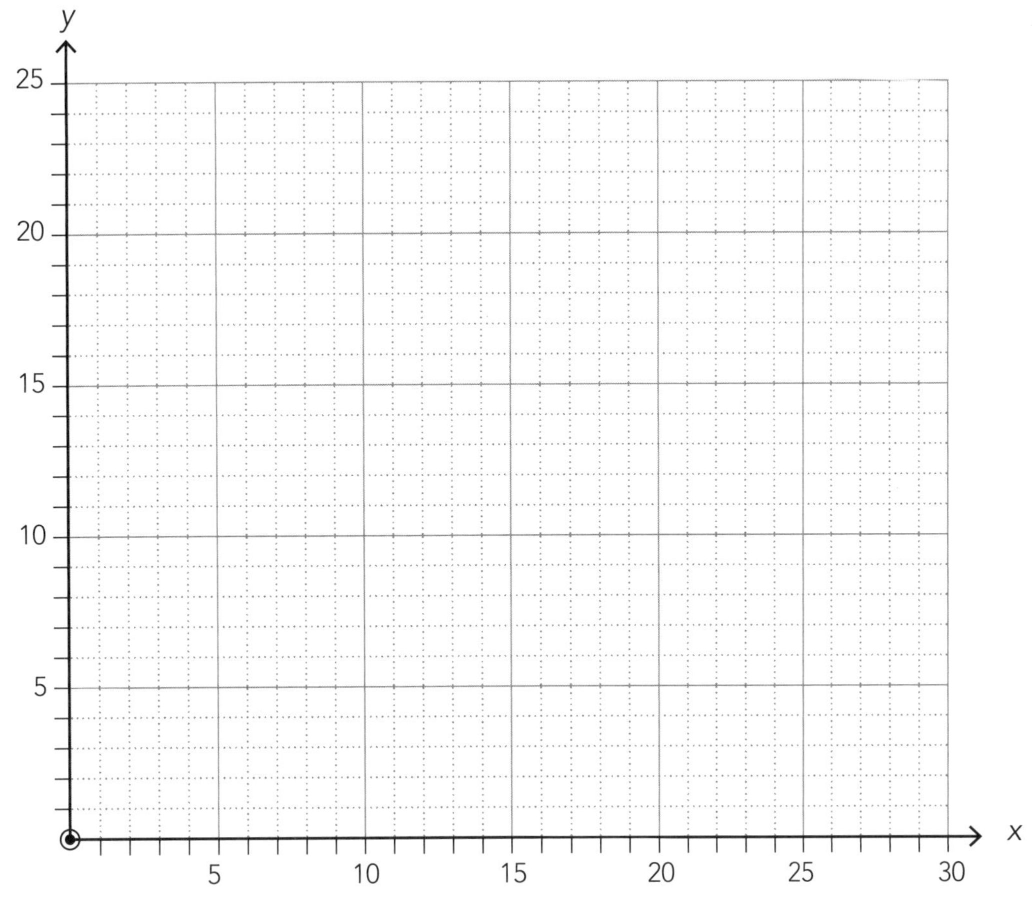

ISBN: 9780170389396

How to deal with vertices that are not whole numbers

These will occur:

- where one or more equations contain < or > symbol (no equals sign)
- where the intercept of two equations is at a point that does **not** have integral (whole number) coordinates.

If either of these occurs, you must consider points that are **close** to the vertex, but still **inside the feasible region**.

Example:

The accommodation for a sports trip is cabins and motels.

- They must hire at least 2 motels in order to have enough cooking facilities.
- The maximum number of motels available is 12.
- There are fewer than 10 cabins available.
- The school policy states that they must hire at least as many cabins as motels.
- There are 15 adults (teachers and parents) available to go on the sports trip, and at least 1 must sleep in each motel and cabin.
- Motels can hold 5 students and cabins can hold 3 students.

Calculate the maximum and minimum number of students that can go on the sports trip.

Variables:

Let x represent the number of motels and y represent the number of cabins.

Objective function: Number of students = $5x + 3y$

Constraints:

$2 \leq x \leq 12$

$y < 10$

$y \geq x$

$x + y \leq 15$

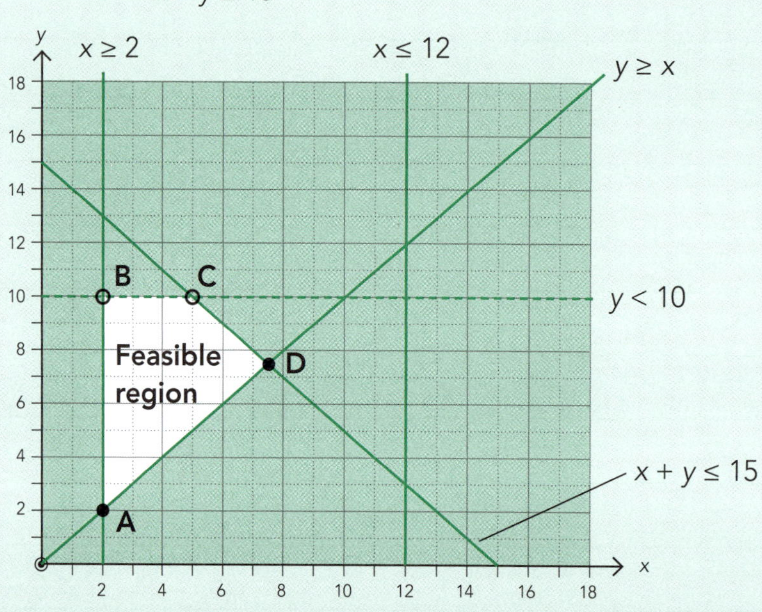

 ISBN: 9780170389396

Vertices of feasible region:

A: (2, 2): This clearly will not produce a maximum number of students.

B: (2, 10): This point can never produce a maximum number of students because the point (5, 10) will always result in more.

C: (5, 10): Not inside the feasible region because $y < 10$.

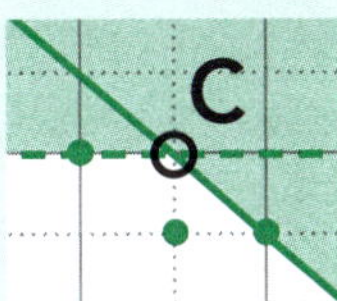

Consider point (4, 10)

Check inequations:	$y < 10$	$10 < 10$	✗

Consider point (5, 9)

Check inequations:	$y < 10$	$9 < 10$	✓
	$x + y \leq 15$	$5 + 9 = 14$	✓

Number of students = 5(5) + 3(9) = **52**

Consider point (6, 9)

Check inequations:	$y < 10$	$9 < 10$	✓
	$x + y \leq 15$	$6 + 9 = 15$	✓

Number of students = 5(6) + 3(9) = **57**

D: (7.5, 7.5) Must have integral points for number of motels and cabins.
∴ consider points nearest to this, but inside the feasible region.

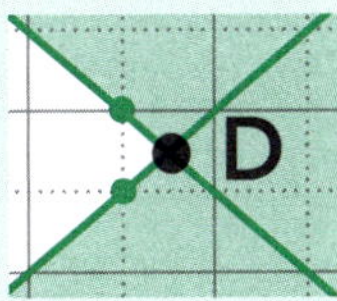

Consider point (7, 7)

Check inequations:	$y \geq x$	$7 \geq 7$	✓
	$x + y \leq 15$	$7 + 7 = 14$	✓

Number of students = 5(7) + 3(7) = **56**

Consider point (7, 8)

Check inequations:	$y \geq x$	$8 \geq 7$	✓
	$x + y \leq 15$	$7 + 8 = 15$	✓

Number of students = 5(7) + 3(8) = **59**

∴ The maximum number of students that can go on the trip is 59, and they will need to hire 7 motels and 8 cabins.

ISBN: 9780170389396

Answer the following questions.

1 Members of the school council are raising funds at their gala day by selling hot dogs and steak sandwiches.

- In their experience, the demand for steak sandwiches is greater than that for hot dogs, so they will make more steak sandwiches than hot dogs.
- They have a maximum of 70 sausages for the hot dogs.
- They have a maximum of 84 pieces of steak for the steak sandwiches.
- Each hot dog requires 1 piece of bread and each steak sandwich requires 2 pieces. They have 190 pieces of bread.

All the ingredients have been donated. If they sell hot dogs for \$3 and steak sandwiches for \$5, how many of each should they make in order to maximise their profit?

ISBN: 9780170389396

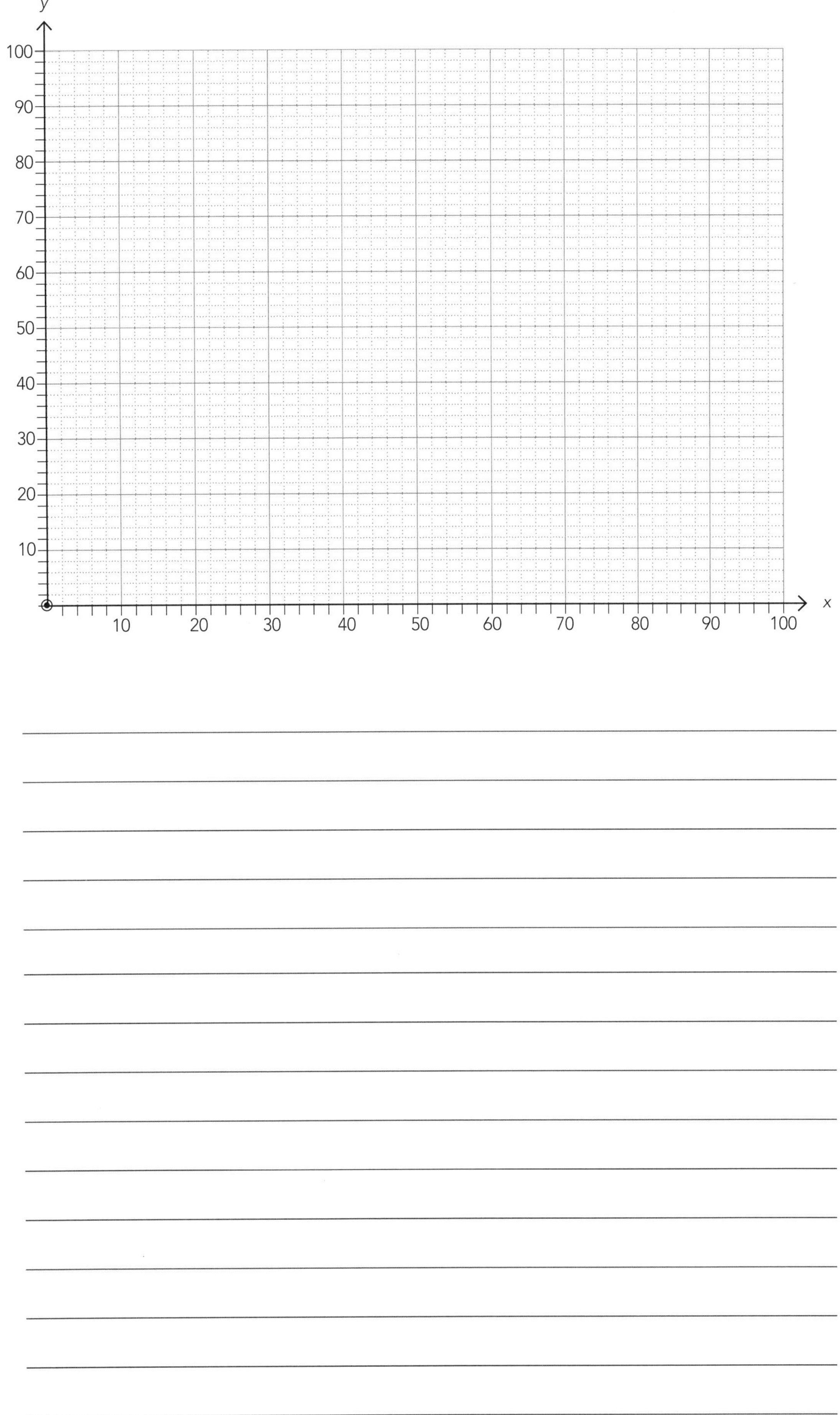

ISBN: 9780170389396

2 A company produces two sizes of garden seats: small and large.

- Small seats need 16 m of timber and large ones need 28 m. The company has 1792 m of timber available.
- Both small seats and large seats need 16 bolts each. There are 1280 bolts available.
- Small seats take 3 hours to make, and large ones need 4 hours. The company wants time use on the seats to be under 270 hours.
- It has firm orders for 6 small seats and 12 large ones.
- The company makes a profit of $30 on small seats and $25 on large seats.

How many of each seat should it make in order to maximise its profit?

  ISBN: 9780170389396

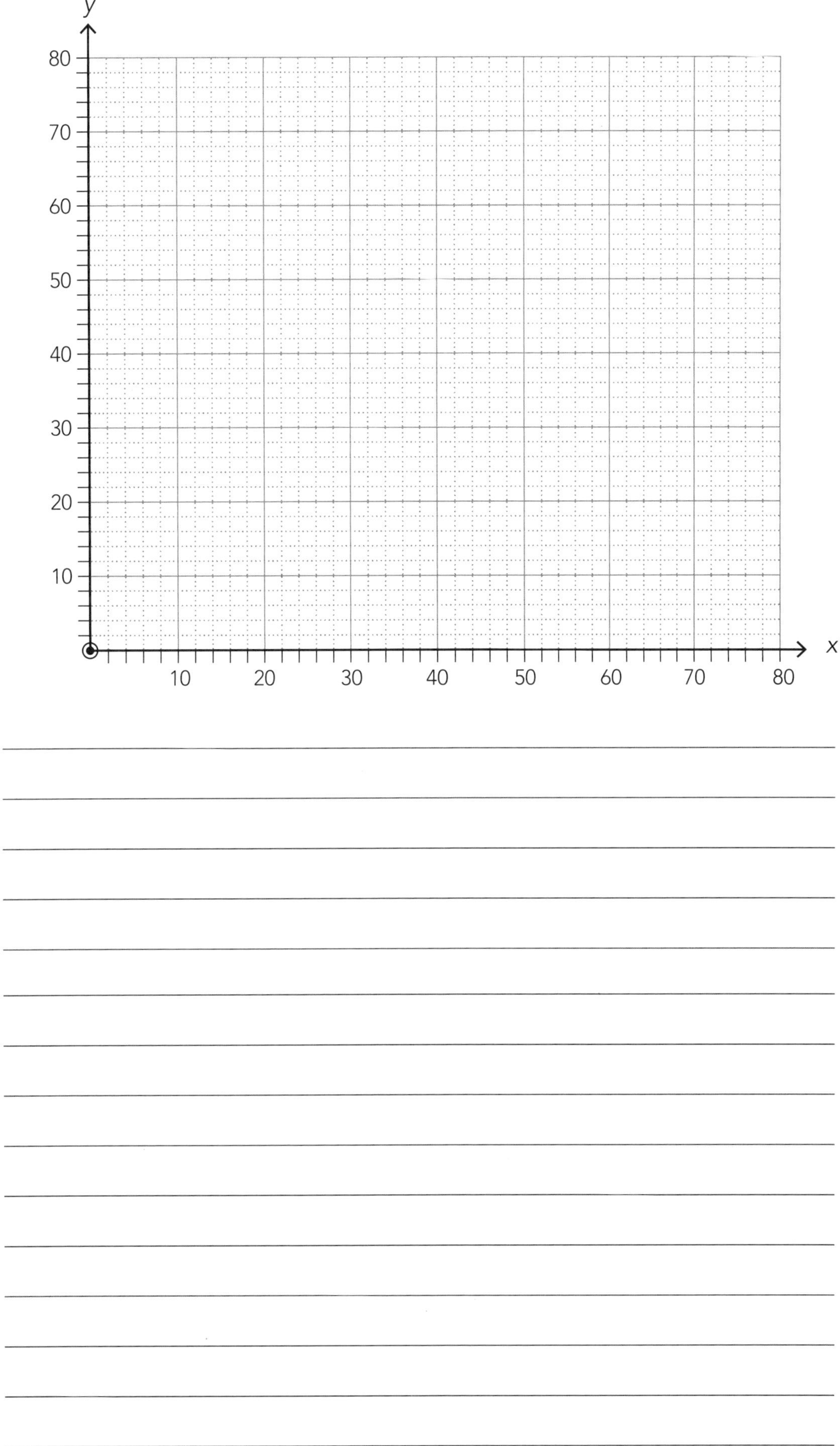
y
80
70
60
50
40
30
20
10
10
20
30
40
50
60
70
80
x

Multiple solutions

- Sometimes two **adjacent** vertices of the feasible region produce the **same** result.
- If this happens, **all valid points between these two vertices** will also produce the **same** result.
- This occurs when the **gradients** of the **objective function** and one of the **constraint** inequations are the **same**.
- You should give **other examples** of values apart from those at the vertices, which also result in the same maximum or minimum.

Example 1:
A pet shop sells dogs and cats.

- It has only 25 cages, so the total number of dogs and cats must not exceed 25.
- From experience, the shop has found that it needs to have at least 5 of each at all times.
- It costs two and a half times as much to feed a dog as it costs to feed a cat. The daily budget for food is $40.
- The shop sells dogs for $75 and cats for $30.

How many dogs and cats should it keep in stock in order to maximise its income?

Variables:
Let x represent the number of dogs and y represent the number of cats.

Objective function: *Income* $= 75x + 30y$

Constraints:
$x + y \leq 25$
$x \geq 5$
$y \geq 5$
$2.5x + y \leq 40$

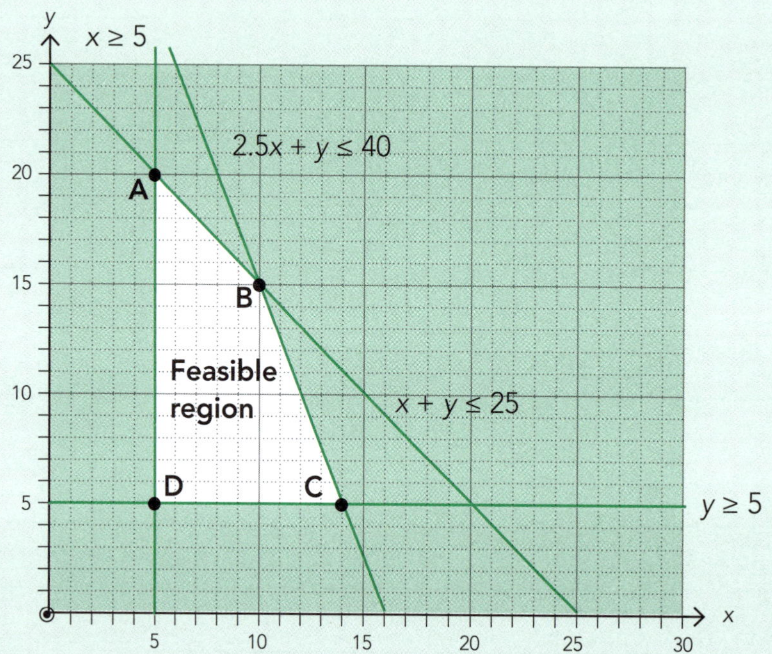

ISBN: 9780170389396

Profit at the vertices of feasible region:

A: (5, 20): $P = 75(5) + 30(20) = 975$

B: (10, 15): $P = 75(10) + 30(15) = 1200$

C: (14, 5): $P = 75(14) + 30(5) = 1200$

Both B (10, 15) and C (14, 5) result in the same profit: $1200.

D: (5, 5): $P = 75(5) + 30(5) = 525$

Because profit of $1200 is obtained at B and C, any **integral** point along the line connecting B and C ($2.5x + y = 40$) will also give the same profit. In this case there is only one other point: (12, 10).

Whole number — some contexts may not require integral answers.

∴ The most profit of $1200 is made at points (10, 15), (12, 10) and (14, 5).

The maximum profit the pet shop can make is $1200, which they get by having 10 dogs and 15 cats,
or 12 dogs and 10 cats,
or 14 dogs and 5 cats.

Explanation for this:

gradient of the objective function = gradient of a constraint inequation

To show this, calculate the gradient of each.

Objective function:

$$Profit = 75x + 30y$$

$$30y = -75x + Profit$$

$$y = -\frac{75}{30}x + \frac{Profit}{30}$$

$$y = -\frac{5}{2}x + \frac{Profit}{30}$$

$$\therefore \text{Gradient} = -\frac{5}{2}$$

Constraint equation:

$$2.5x + y = 40$$

$$y = -2.5x + 40$$

$$\therefore \text{Gradient} = -2.5 = -\frac{5}{2}$$

You may be asked to use this concept in another way.

Example 2:

A pet shop also sells rats and mice.

- Health and safety regulations demand that the total number of rats and mice does not exceed 50.
- The shop has cages that can hold a maximum of 32 rats and 40 mice.
- It likes to have at least 8 rats and 10 mice at all times.
- It costs one and a half times as much to feed and care for a rat as it costs to feed and care for a mouse. The daily budget for feeding and caring for the rats and mice is $60.
- The shop sells mice for $5.

a What price should the shop charge for rats if its maximum income is the same for selling between 20 and 32 rats inclusive?

b Find the value of the maximum income and confirm that it occurs only when the number of rats sold is between 20 and 32 inclusive.

c List all the combinations of sales of rats and mice that would produce it.

Variables:

Let x represent the number of rats and y represent the number of mice.

Objective function: $Income = ax + 5y$

Constraints:

$x + y \leq 50$

$8 \leq x \leq 32$

$10 \leq y \leq 40$

$3x + 2y \leq 120$

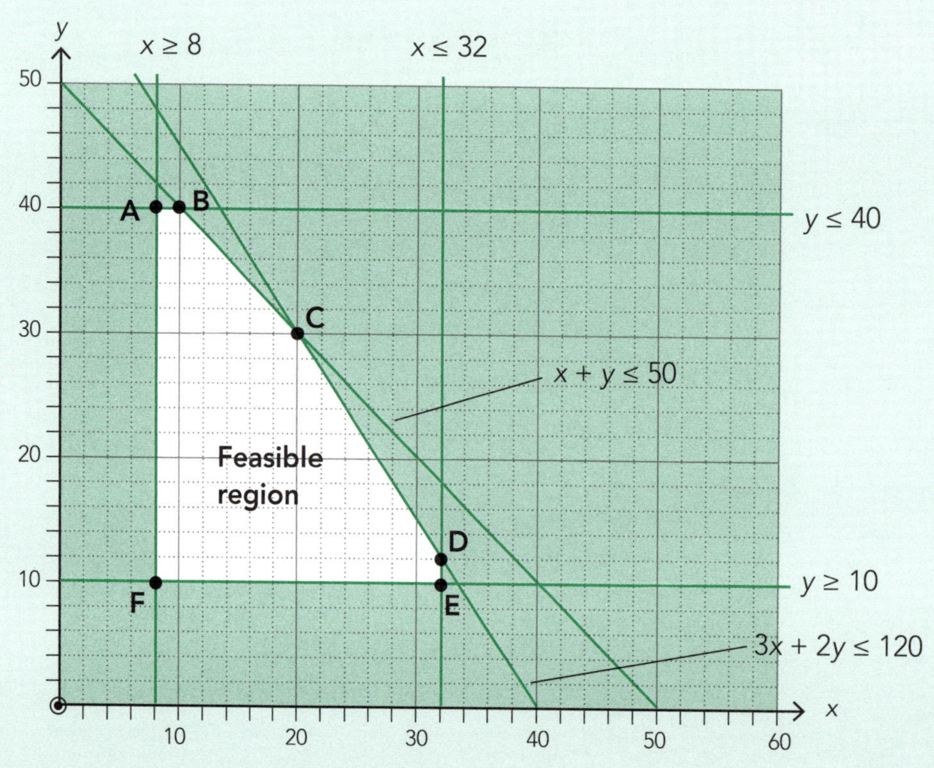

ISBN: 9780170389396

a Maximum income from selling between 20 and 32 rats is between C and D.

$$\therefore 3x + 2y \leq 120.$$

Income is the same along C and D if the gradient of the objective function matches that of CD.

Gradient of CD:

$$2y = -3x + 120$$
$$y = -\frac{3}{2}x + 60$$
$$m = -\frac{3}{2}$$

Gradient of objective function:

$$Income = ax + 5y$$
$$5y = -ax + Income$$
$$y = -\frac{a}{5}x + \frac{Income}{5}$$
$$m = -\frac{a}{5}$$

Gradient of CD = Gradient of objective function $\Rightarrow -\frac{a}{5} = -\frac{3}{2}$

$$2a = 15$$
$$a = 7.5$$

∴ They would have to sell rats for $7.50 for the income at C and D to be the same.

b Vertices that could produce the maximum income:

C: (20, 30): $P = 7.5(20) + 5(30) = 300$
D: (32, 12): $P = 7.5(32) + 5(12) = 300$

∴ The maximum income of $300 occurs at points C and D, at which 20 rats and 30 mice, or 32 rats and 12 mice are sold.

c All the combinations between C and D that produce an income of $300:

(20, 30): $P = 7.5(20) + 5(30) = 300$ ⇒ 20 rats and 30 mice
(24, 24): $P = 7.5(24) + 5(24) = 300$ ⇒ 24 rats and 24 mice
(28, 18): $P = 7.5(28) + 5(18) = 300$ ⇒ 28 rats and 18 mice
(32, 12): $P = 7.5(32) + 5(12) = 300$ ⇒ 32 rats and 12 mice

Note: This question could be asked the other way round. You could be told the price of both rats and mice, and then asked to find the gradient of the line between two points that produce the same income.

Answer the following questions.

1 The canteen is making mini-pizzas and stuffed potatoes.

- They have room for a total of 40 items in their oven.
- They have ingredients for up to 20 stuffed potatoes and up to 30 mini-pizzas.
- It takes 3 minutes to make each stuffed potato and 1 minute to make each mini-pizza. The staff has a total of 72 minutes in which to get the mini-pizzas and stuffed potatoes made.
- The canteen makes a profit of 50 cents on each mini-pizza and 75 cents on each stuffed potato.
- They sell everything they make.

a Calculate the numbers of stuffed potatoes and mini-pizzas that the canteen should make in order to maximise its profit.

ISBN: 9780170389396

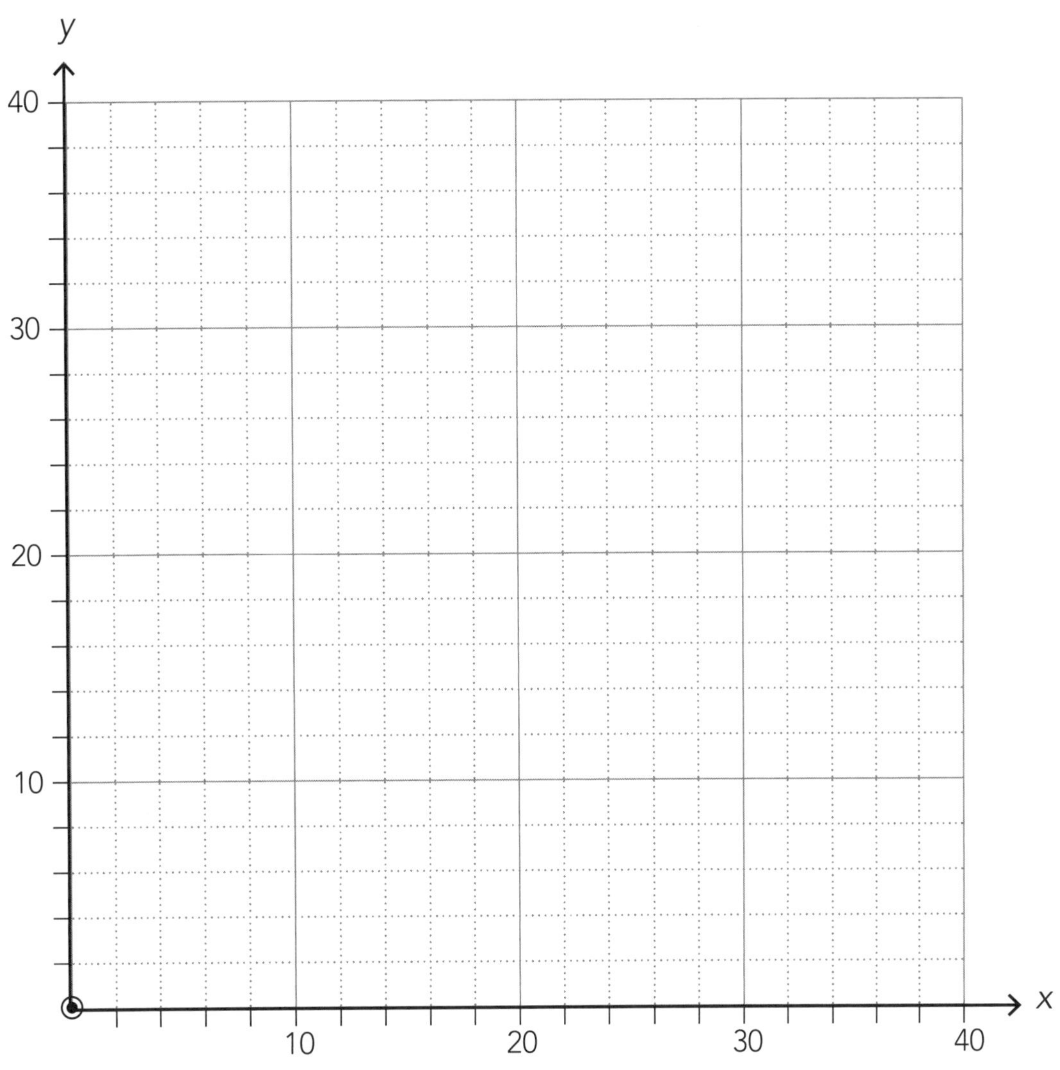

b Investigate what would happen to the total profit of the canteen if they reduced the profit from mini-pizzas from 50 cents to 25 cents.

c By how much would they have to increase their oven capacity for it to no longer be a constraint on the number of mini-pizzas and stuffed potatoes made? Assuming they left the profits on mini-pizzas and stuffed potatoes at 50 cents and 75 cents, how would their profit change?

d The canteen management doesn't want to increase the oven capacity, but they might change the profits. They want to leave the profit from stuffed potatoes at 75 cents but would consider increasing the profit from mini-pizzas. What would the profit from each mini-pizza need to be, if the maximum overall profit from both products was the same for points C and D? Are there other points which would generate the same profit?

ISBN: 9780170389396

2 Eru assembles electronics kits in his spare time.

- He has enough boxes to make up 50 kits.
- He has enough components to assemble 40 beginner's kits and 30 intermediate kits.
- Each beginner's kit takes him 2 minutes to assemble, and each intermediate kit takes him 3 minutes. He has 2 hours available to assemble the kits.
- He makes a profit of $5 on each beginner's kit and $6 on each intermediate kit.

a How many of each should he make in order to maximise his profit?

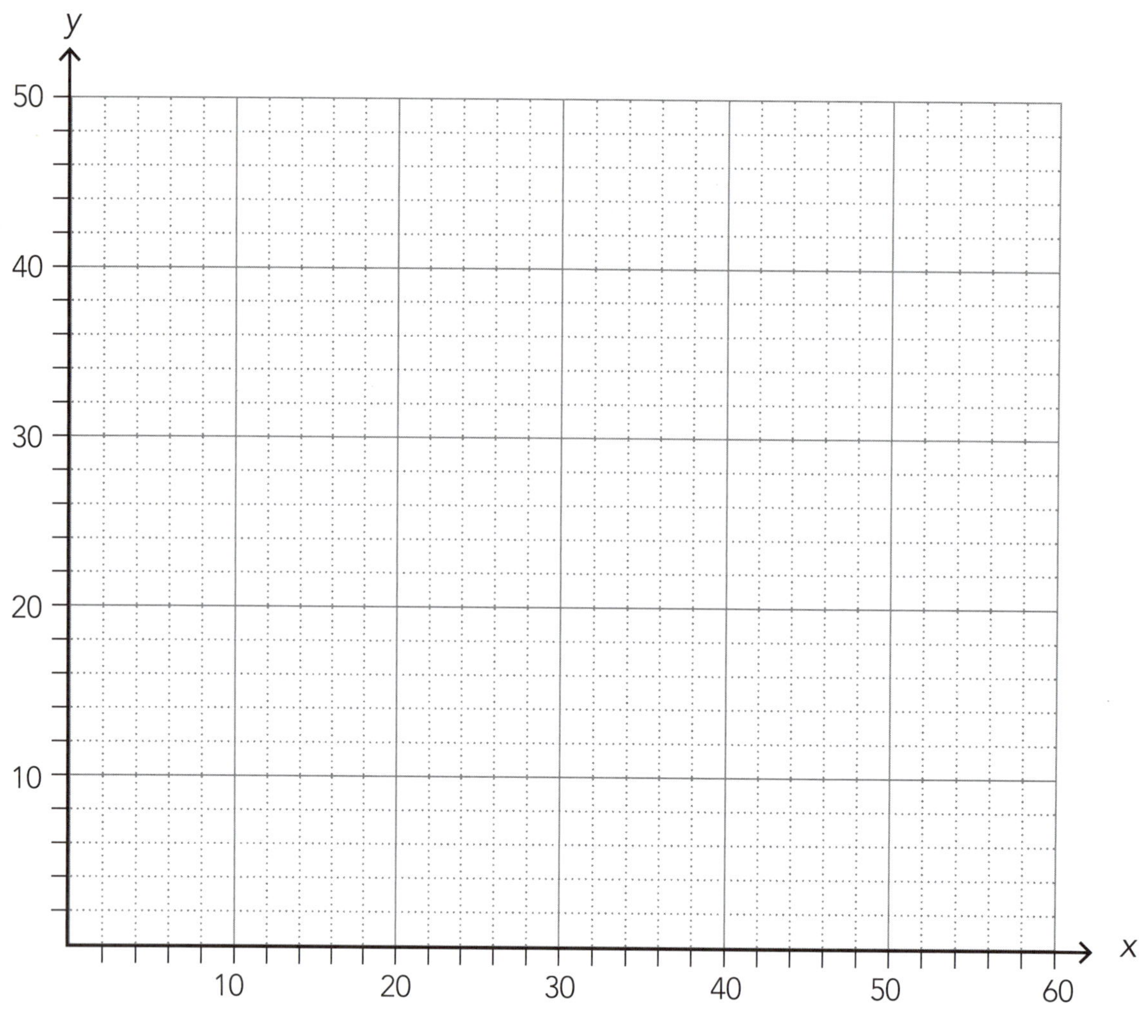

b Investigate what would happen to Eru's overall profit if he increased the profit he made on the beginner's kit to $6.

 ISBN: 9780170389396

c By how much would Eru have to increase his time for assembling the kits, for time to no longer be a constraint on the number of kits he makes up? Assuming that he leaves the profits on beginner and intermediate kits at $5 and $6, how would his profit change?

d His accountant suggests that he shouldn't increase his time, but he could increase his profit from beginner kits to $6. What would the profit on intermediate kits need to be for his maximum overall profit to be the same between (15, 30) and (30, 20)? What would his new maximum profit be? Are there other points which would generate the same profit?

Practice tasks

Practice task one

Pania is training for a swimming competition. She would like to enter several freestyle and backstroke events, and wants to improve her times.

This activity requires you to use linear programming to model the constraints Pania has on the number of 50-metre lengths she swims using each stroke. You need to make recommendations about the optimum number of each length she should swim in order to maximise her return from sponsors. You need to discuss the effect of changing the instructions given to her by her coach.

You will present your findings supported by graphs, equations and relevant calculations.

Information:

- Her coach insists that she completes at least 36 lengths in total, with at least 12 of those being backstroke.
- Because he feels she has more potential to improve in freestyle, he wants her to do at least 2 freestyle lengths for every backstroke length completed.
- She takes an average time of 35 seconds for a length of backstroke and 30 seconds for a length of freestyle. She needs to complete the training session in 35 minutes or less.
- She has sponsorship of 75 cents per backstroke length and 55 cents per freestyle length.
- Her coach changes his mind, and decides that he wants her to do at least 3 backstroke lengths for every freestyle length completed.

 ISBN: 9780170389396

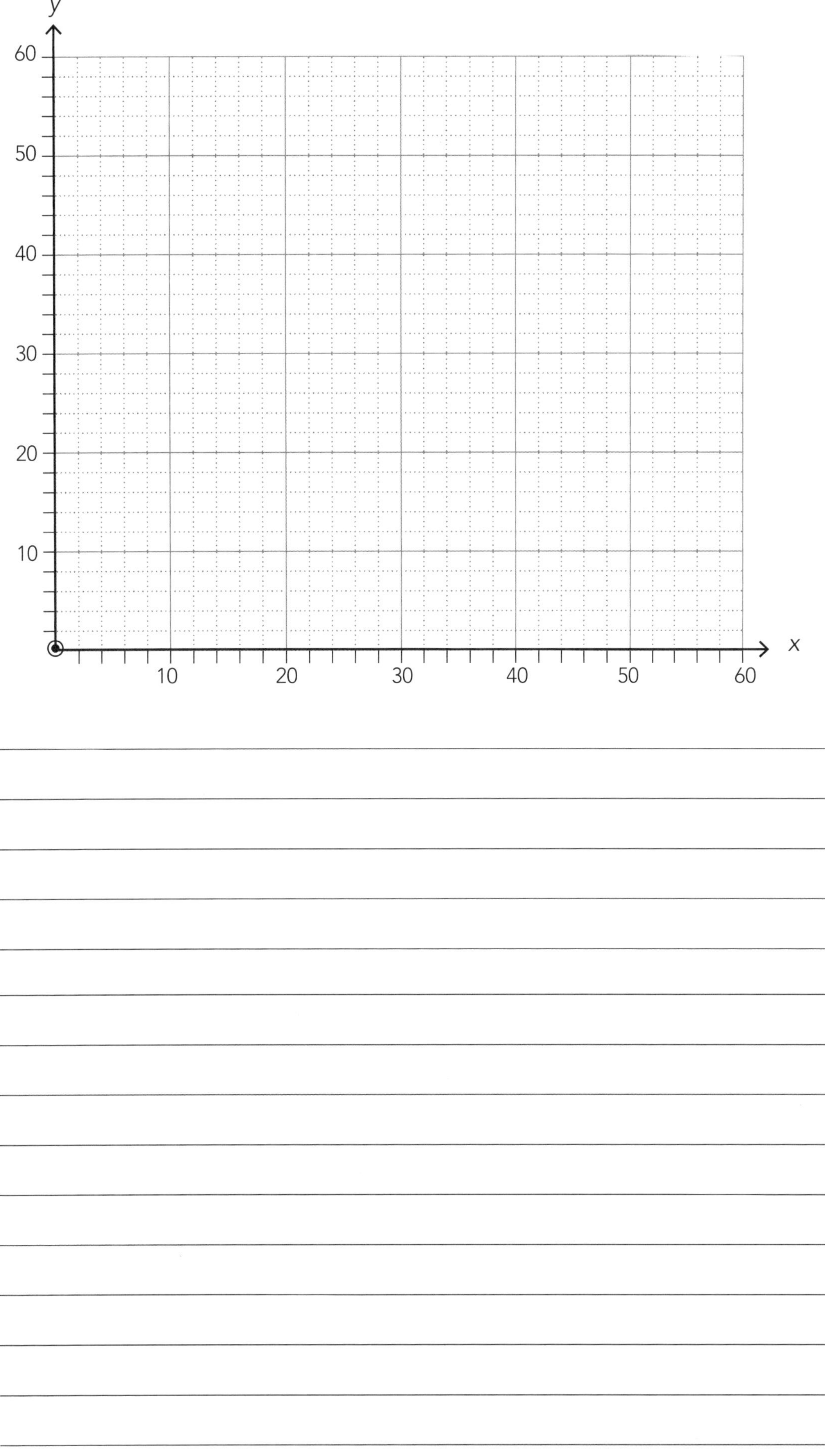
y
60
50
40
30
20
10
10
20
30
40
50
60
x

Practice task two

Farmer White raises hens and ducks and he sells their eggs.

This activity requires you to use linear programming to model the constraints Farmer White has for the number of hen eggs and duck eggs he should produce. You need to make recommendations about the optimum number of each type of egg he should produce in order to maximise his income. You need to discuss the effect of increasing the price of duck eggs and increasing his budget for chicks and ducklings on his income.

You will present your findings supported by graphs, equations and relevant calculations.

Information:

- All his hens and ducks lay one egg per day.
- The birds are all free range, but he has enough night-time shelters for just 100 birds.
- He keeps a maximum of 80 hens.
- Customer demand means that he must produce four or fewer duck eggs for every hen egg produced.
- Chicks cost $1 each and ducklings cost $2. The maximum in his budget for chicks and ducklings is $144.
- He sells hen eggs for 60 cents each and duck eggs for $1.00 each.
- He is considering increasing the price of duck eggs to $1.20.
- As well as increasing the price of duck eggs to $1.20, he is also considering increasing the amount in his budget for chicks and ducklings. He would like to know how much he has to increase it by for it to no longer be a constraint on the number of eggs produced.

ISBN: 9780170389396

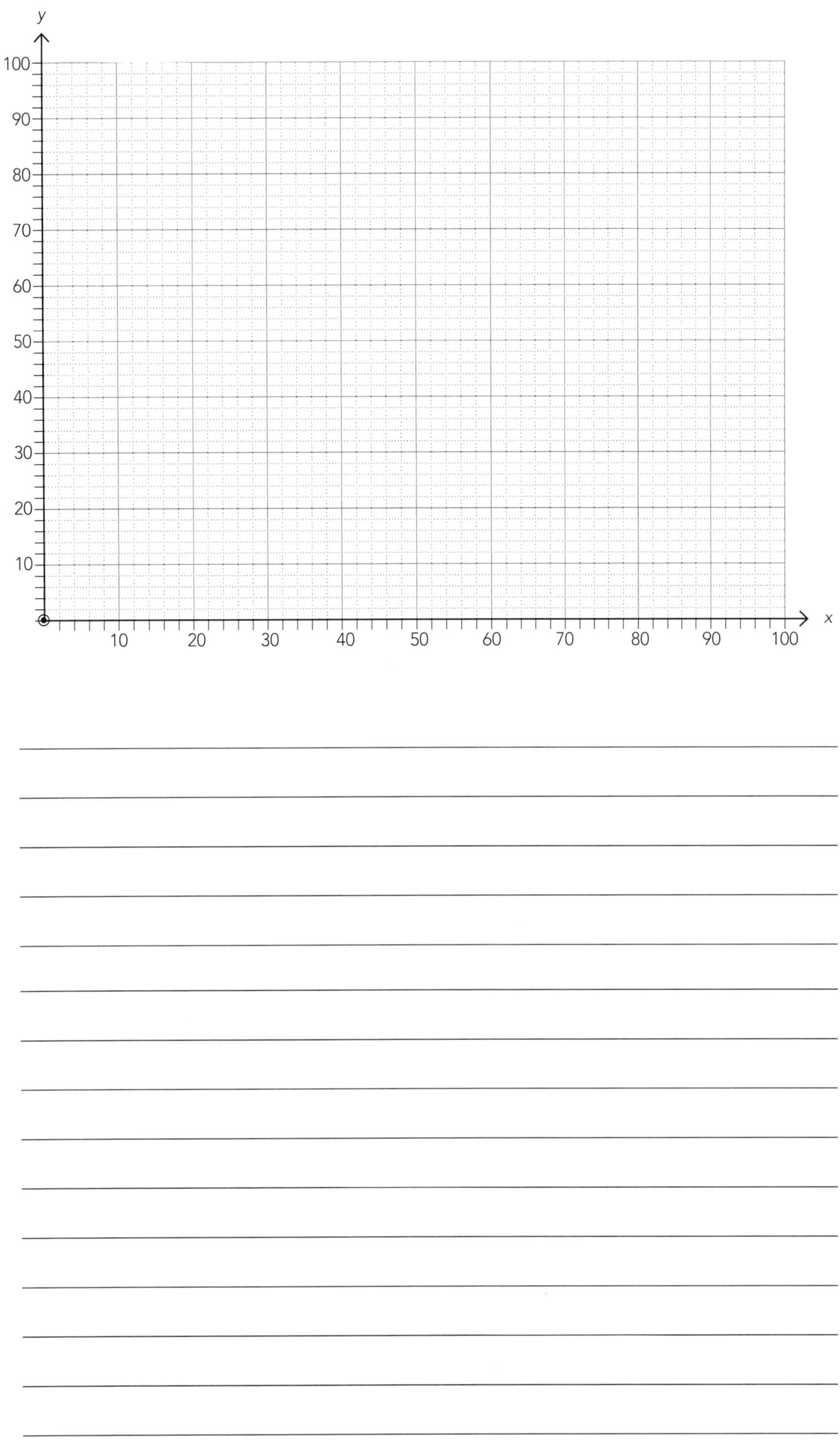

ISBN: 9780170389396

Practice task three

A farmer is mixing food for his pigs.

This activity requires you to use linear programming to model the constraints the farmer has for the number of bags of ingredient A and ingredient B he should use when mixing his pig food. You need to make recommendations about the optimum number of each type of ingredient he should use in order to minimise his cost of a **batch** of feed and each **bag** of feed. You need to discuss the effect of changing the protein requirement on his cost of feed.

You will present your findings supported by graphs, equations and relevant calculations.

- Ingredient A contains 3 units of carbohydrate and ingredient B contains 2 units. The food must contain at least 36 units of carbohydrate in total.
- Ingredient A contains 1 unit of protein and ingredient B contains 2 units. The food must contain at least 20 units of protein in total.
- He must include at least 4 bags of ingredient A and at least 2 bags of ingredient B in each batch.
- Once the feed is mixed, he stores it in the ingredient bags.
- Ingredient A costs $11 per bag and ingredient B costs $18 per bag. He must make up his mix with entire bags because the ingredients don't keep once the bag has been opened.
- Calculate the number of bags of ingredients A and B that should go into his mix in order to minimise the price of one **batch** of feed.
- Calculate the number of bags of ingredients A and B that should go into his mix in order to minimise the price of one **bag** of feed.
- He shows his formulation to his vet, and the vet suggests that he increase the protein content to be more than 20 units.

  ISBN: 9780170389396

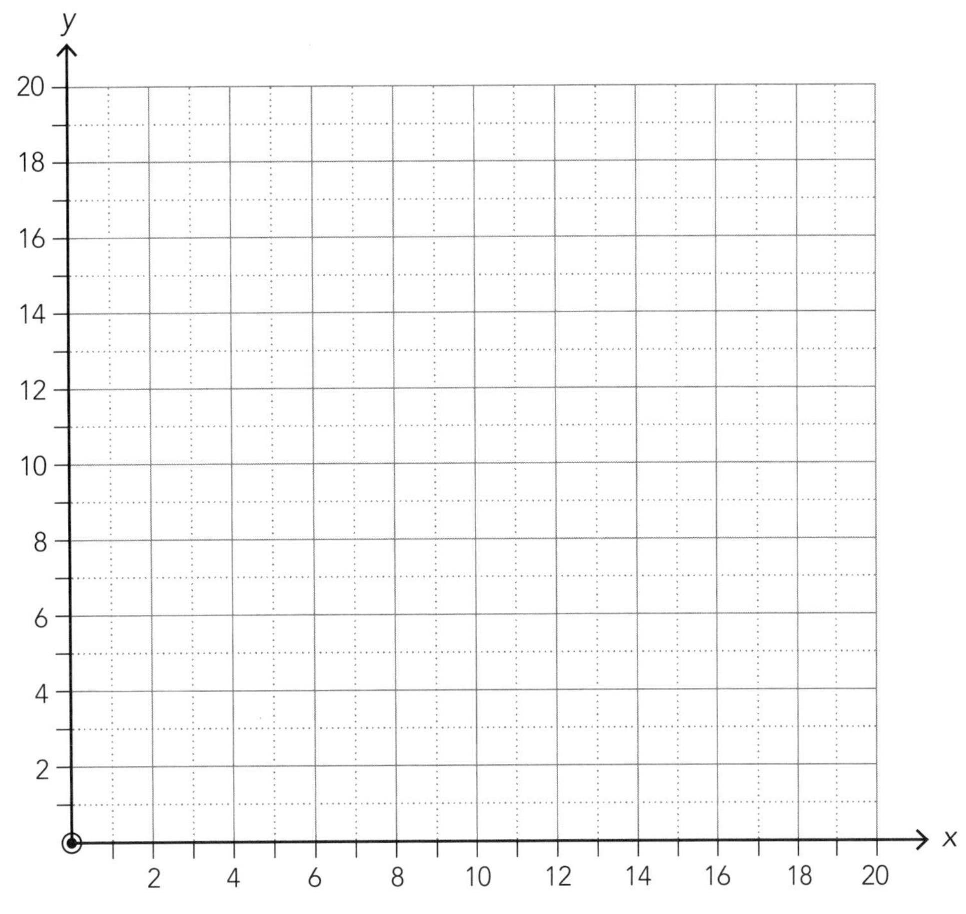
y
20
18
16
14
12
10
8
6
4
2
2
4
6
8
10
12
14
16
18
20
x

Answers

Drawing straight lines (pp. 5–18)

1 Plotting points using the equation (pp. 5–7)

1 **2**

3 **4&5**

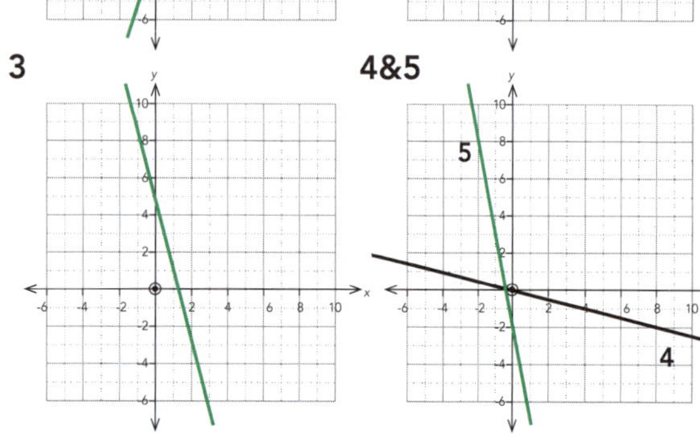

6&7

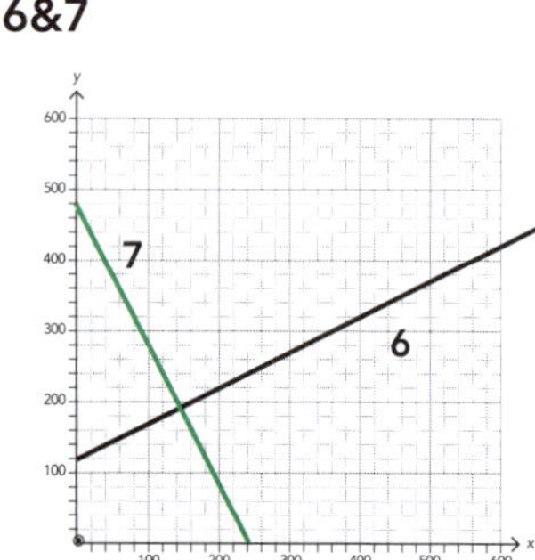

2 Plotting points using the table function on your calculator (pp. 8–10)

1&2 **3&4**

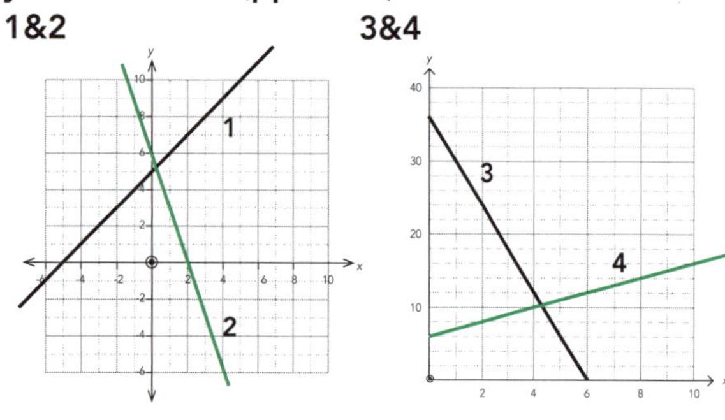

5&6 **7&8**

9&10 **11&12**

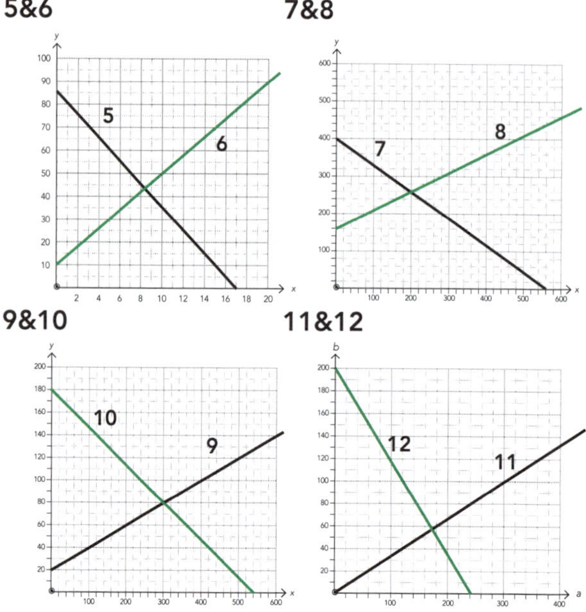

3 Using the x and y intercepts (pp. 11–13)

1&2 **3&4**

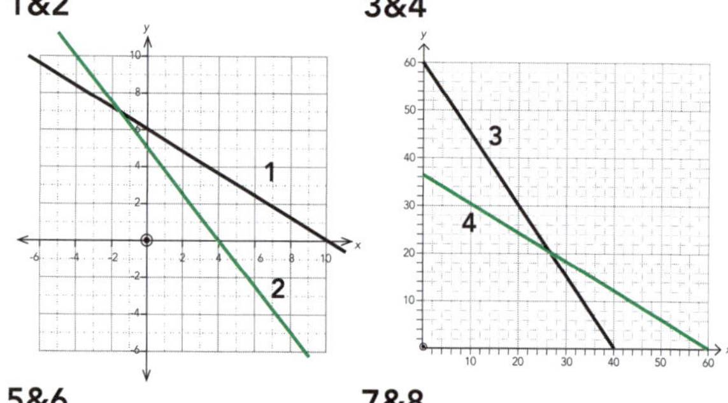

5&6 **7&8**

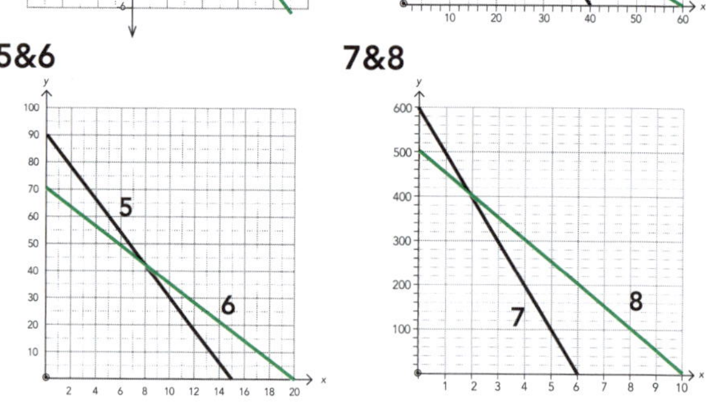

9&10 **11&12**

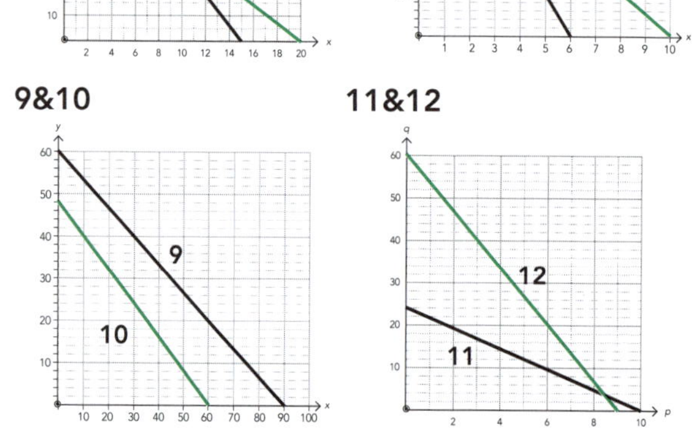

 ISBN: 9780170389396

4 Horizontal and vertical lines (pp. 14–15)

1

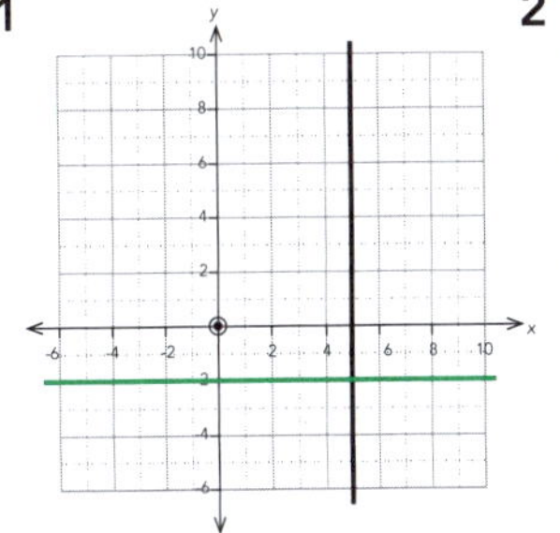

2

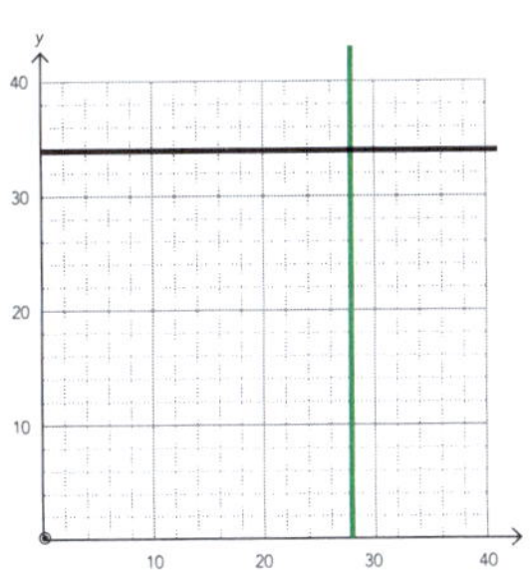

3

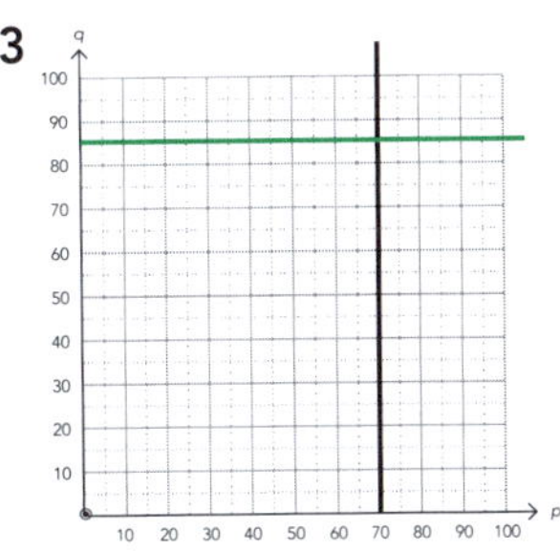

Test yourself (pp. 17–18)

1&2

3&4

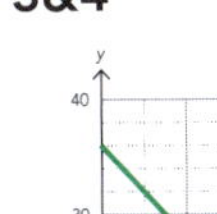

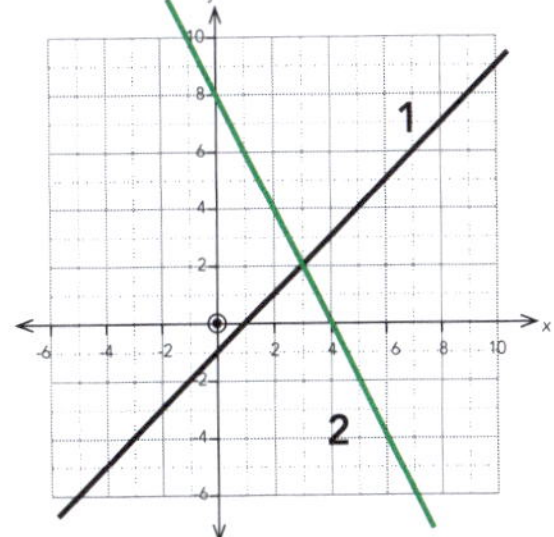

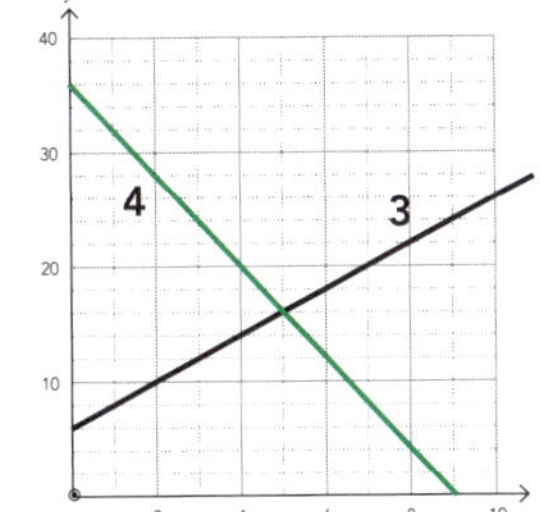

5&6

7&8

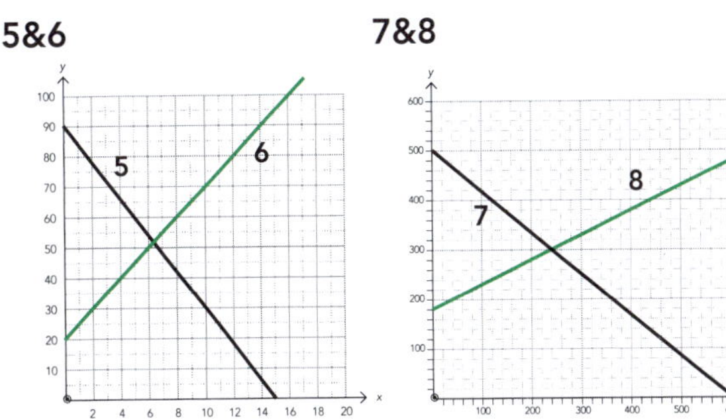

9&10

11&12

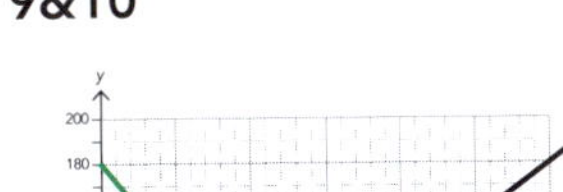

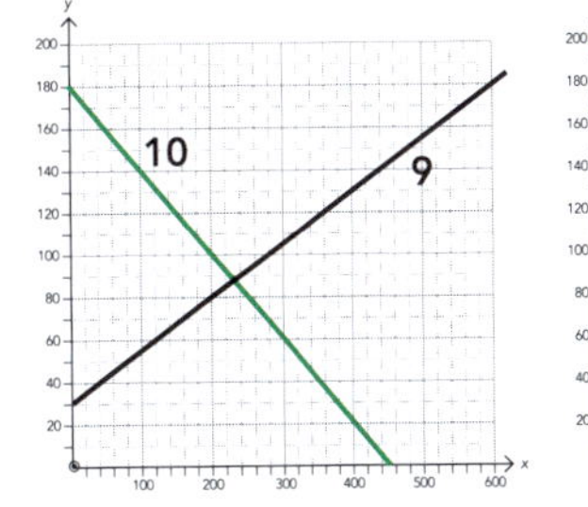

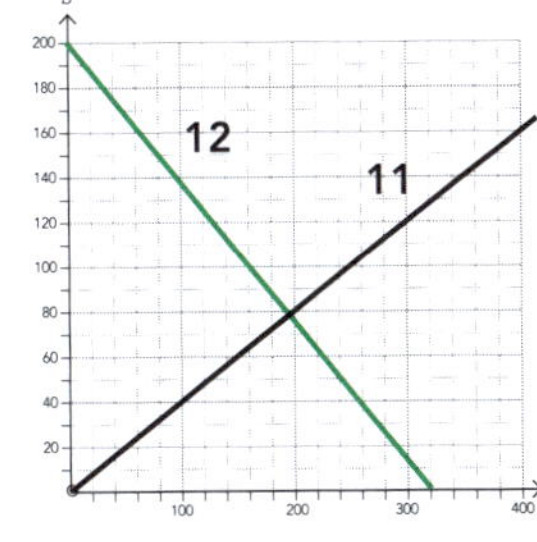

Finding where lines meet (intersections) (pp. 19–28)

1 By drawing the lines (pp. 19–22)

1

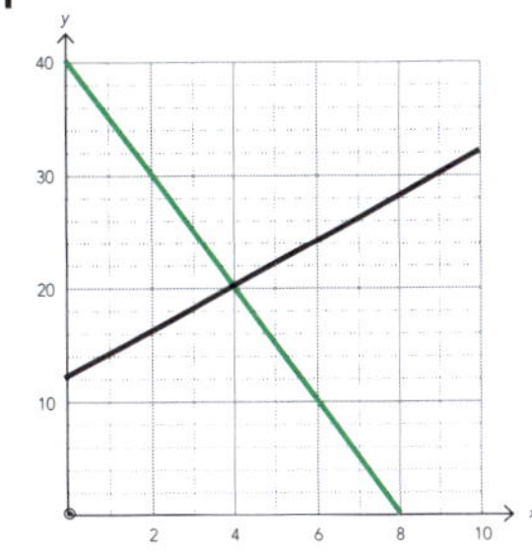

Solution: (4, 20)

2

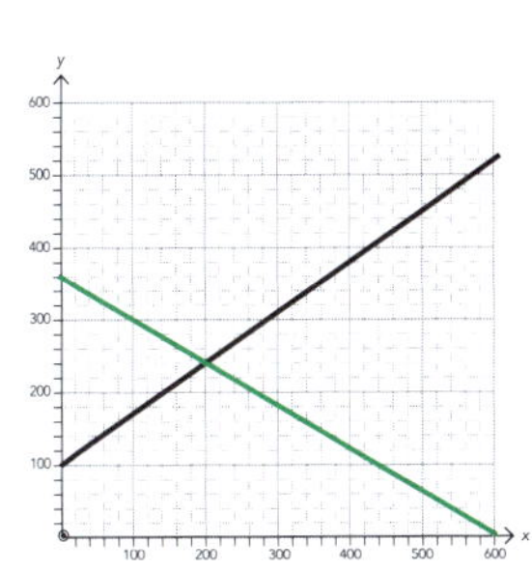

Solution: (200, 240)

3

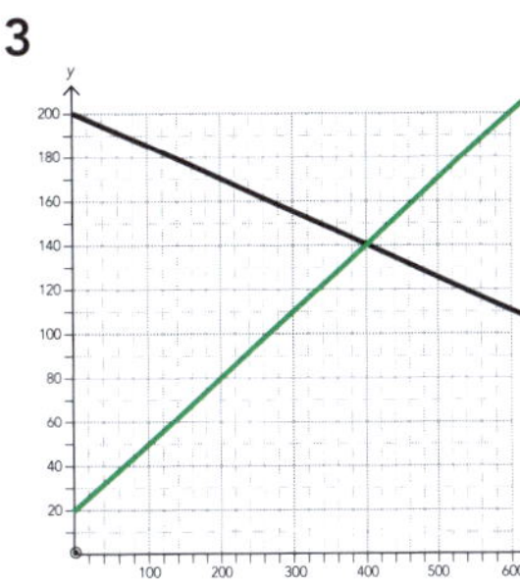

Solution: (400, 140)

4

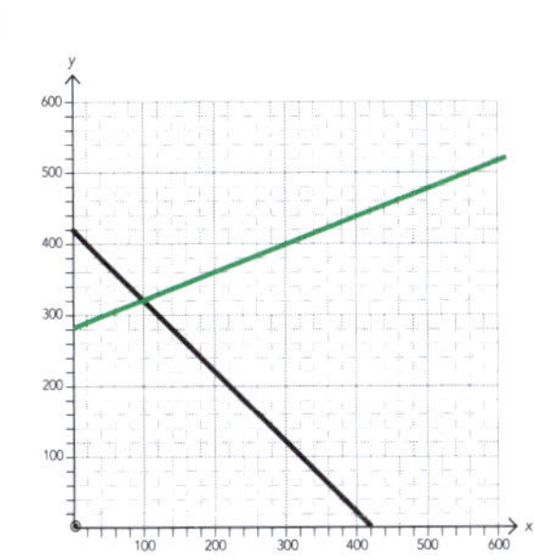

Solution: (100, 320)

5

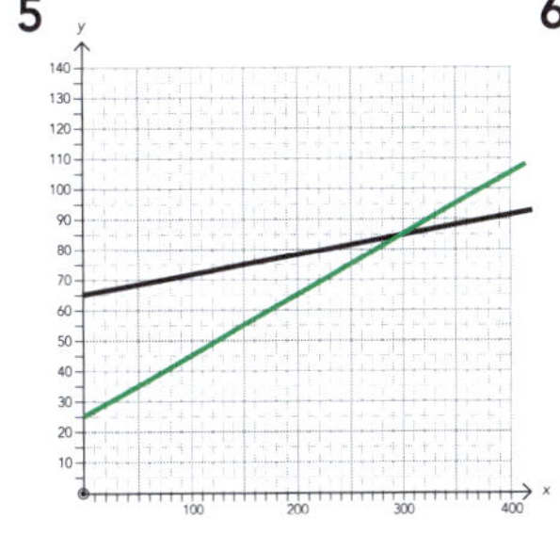

Solution: (300, 85)

6

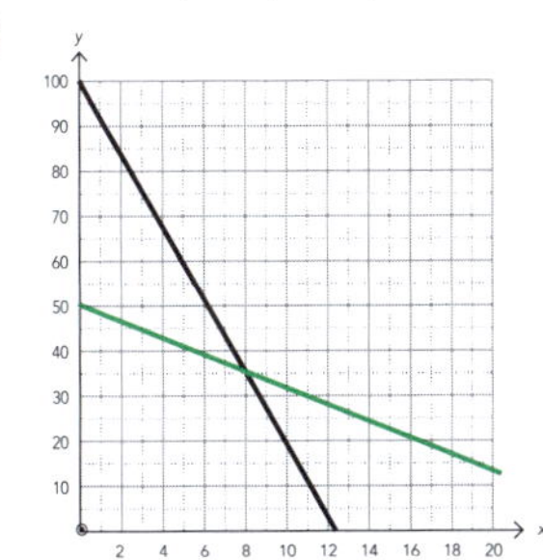

Solution: (8, 36)

Test yourself (pp. 27–28)

1	(-3, 9)	**2**	(-1, 6)
3	(11, 20)	**4**	(7, 11)
5	(9, 7)	**6**	(22, 27)
7	(4, 20)	**8**	(8, 36)
9	(29, 1.5)	**10**	(21.25, 16.75)
11	(11, 20)	**12**	(25, 18)
13	(9, 7)	**14**	(28, 32)
15	(10, 17)	**16**	(4.5, 6)
17	(11, 13)	**18**	(10.2, 11.8)
19	(43, 29)	**20**	(50.2, 81.5)

ISBN: 9780170389396

Inequations (pp. 29–45)

Understanding the words (pp. 29–32)

1

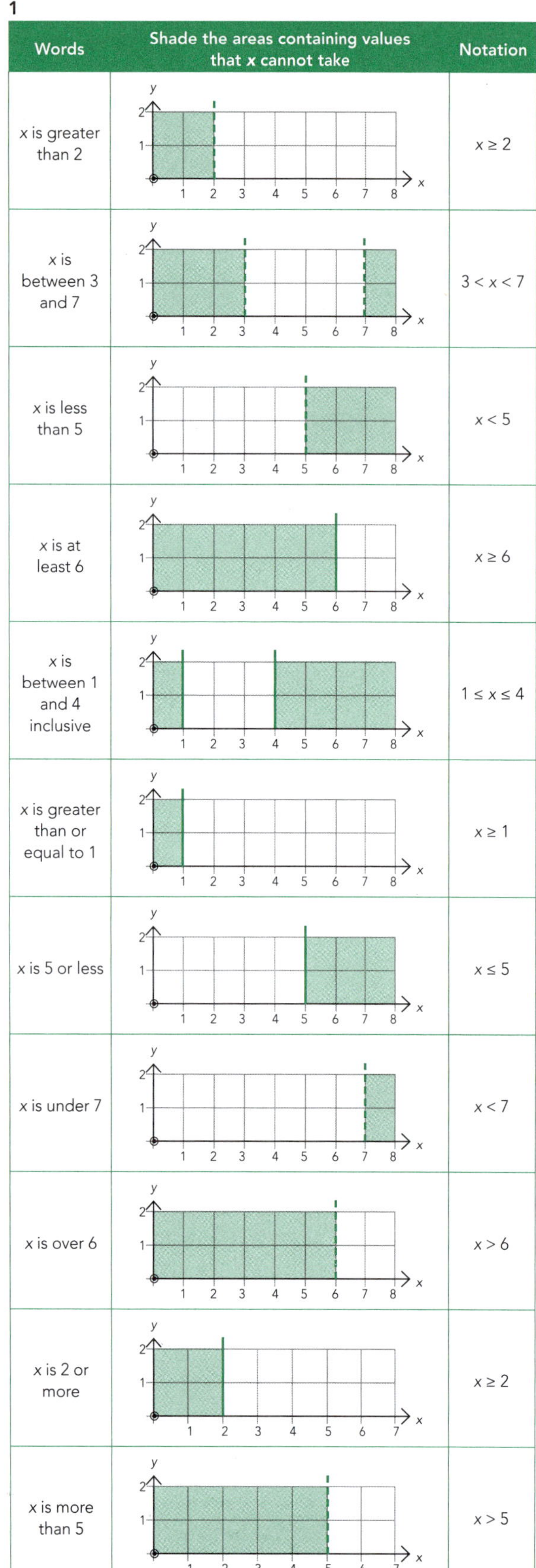

Words	Shade the areas containing values that ***x*** cannot take	Notation
x is greater than 2		$x \geq 2$
x is between 3 and 7		$3 < x < 7$
x is less than 5		$x < 5$
x is at least 6		$x \geq 6$
x is between 1 and 4 inclusive		$1 \leq x \leq 4$
x is greater than or equal to 1		$x \geq 1$
x is 5 or less		$x \leq 5$
x is under 7		$x < 7$
x is over 6		$x > 6$
x is 2 or more		$x \geq 2$
x is more than 5		$x > 5$
x is not more than 3		$x \leq 3$
x exceeds 4		$x > 4$
x is not less than 7		$x \geq 7$

2

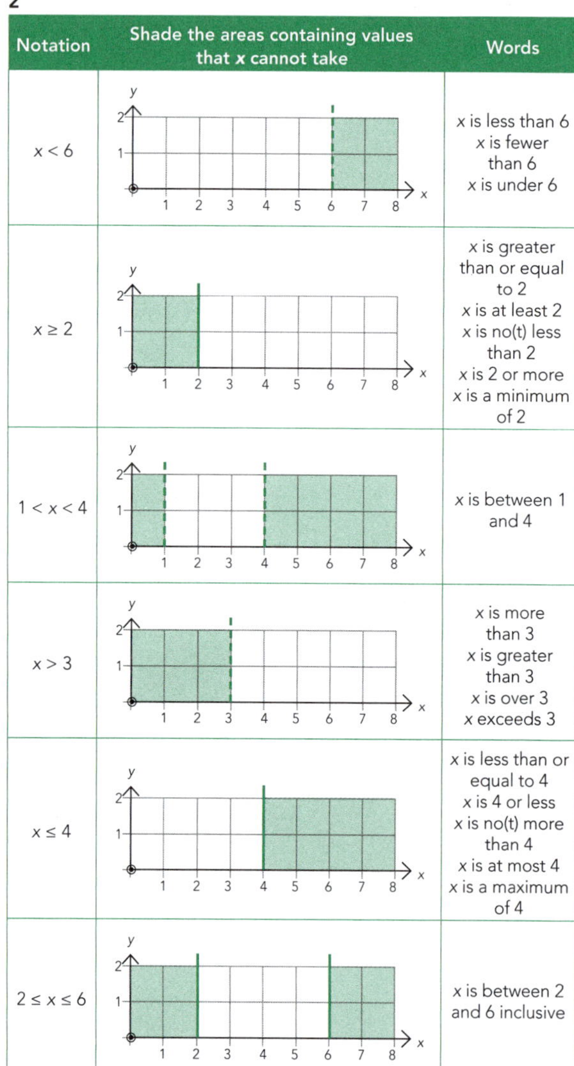

Notation	Shade the areas containing values that ***x*** cannot take	Words
$x < 6$		*x* is less than 6 *x* is fewer than 6 *x* is under 6
$x \geq 2$		*x* is greater than or equal to 2 *x* is at least 2 *x* is no(t) less than 2 *x* is 2 or more *x* is a minimum of 2
$1 < x < 4$		*x* is between 1 and 4
$x > 3$		*x* is more than 3 *x* is greater than 3 *x* is over 3 *x* exceeds 3
$x \leq 4$		*x* is less than or equal to 4 *x* is 4 or less *x* is no(t) more than 4 *x* is at most 4 *x* is a maximum of 4
$2 \leq x \leq 6$		*x* is between 2 and 6 inclusive

ISBN: 9780170389396

Rearranging inequations (pp. 33–34)

1 $y \geq -2.5x + 15$ **2** $5x - 2y < -2$

3 $y \geq x + 0$ **4** $4x - y \leq 12$

5 $y < 3x + 10$ **6** $9x - 2y > -36$

7 $y > \frac{1}{5}x - \frac{6}{5}$ **8** $4x - y \geq 11$

9 $y < -4x + 17$ **10** $95x - 20y \leq 100$

Drawing inequations (pp. 35–43)

1 **a** H **b** F
c E **d** B
e C **f** G
g D **h** A

2 **a**

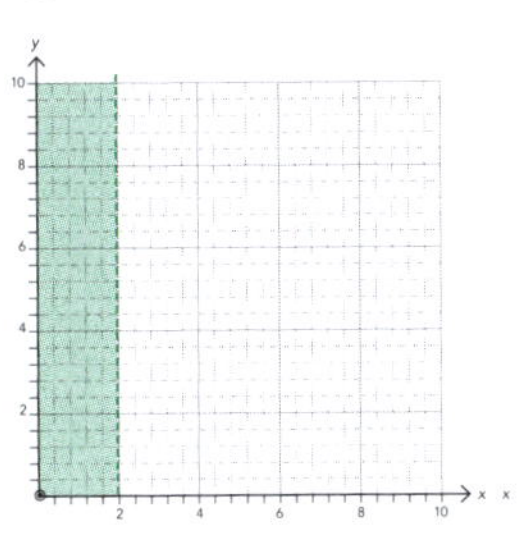

b

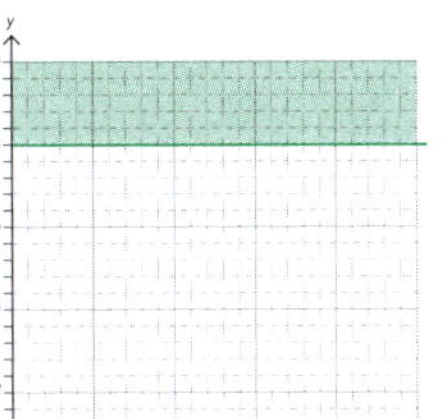

c

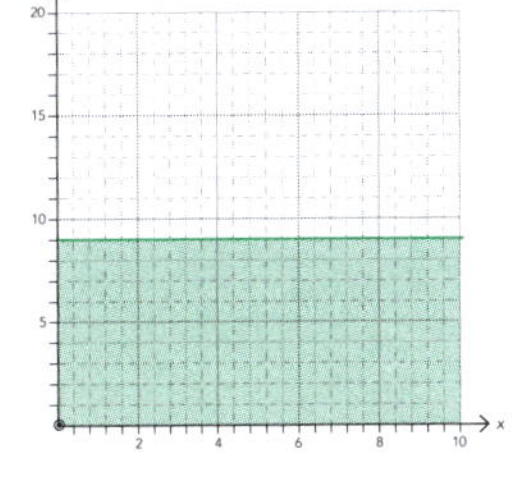

d

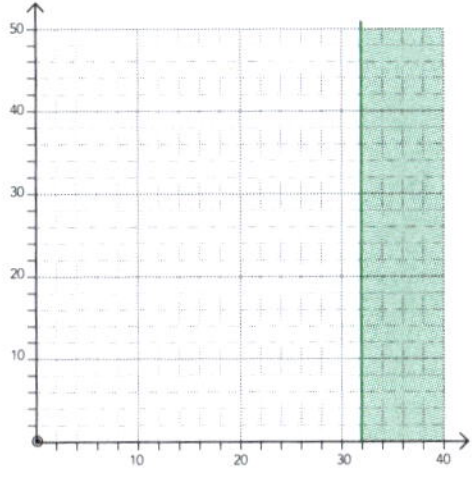

e

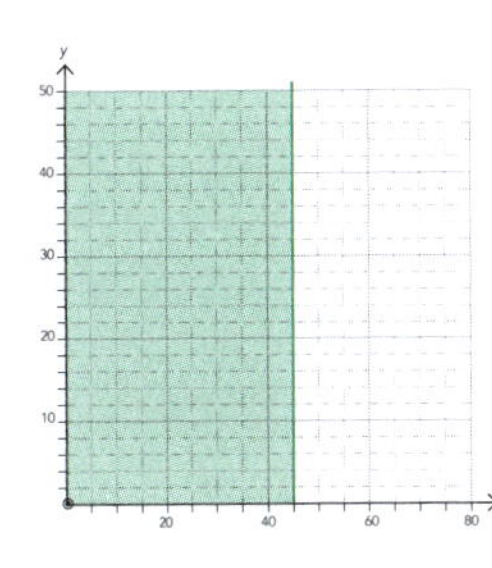

f

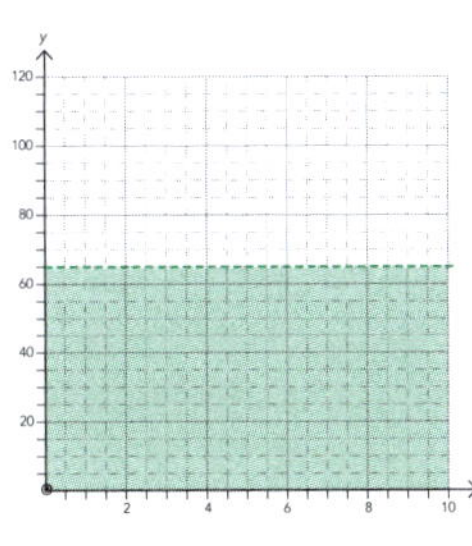

3 **a**

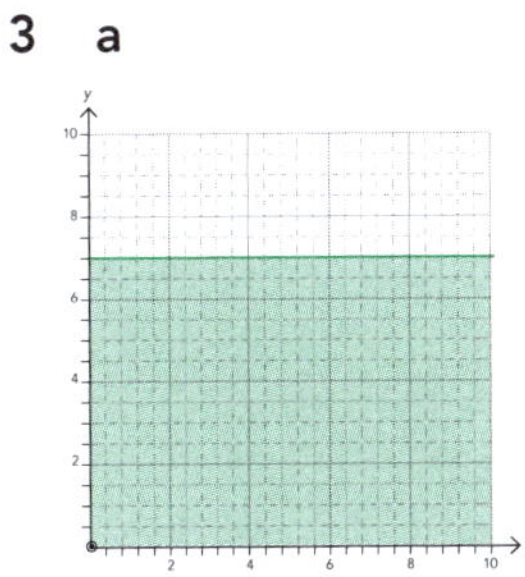

b

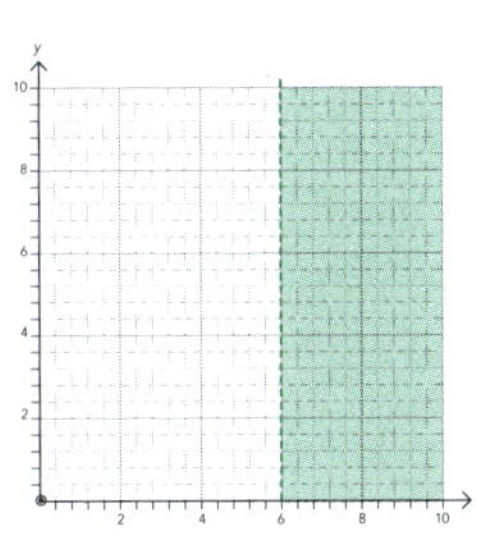

c

d

e

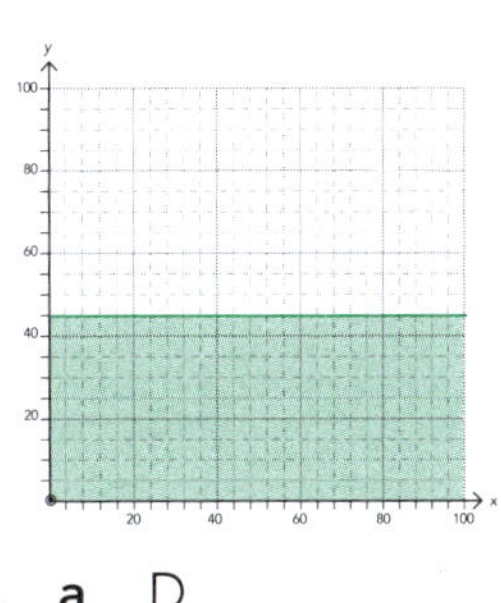

f

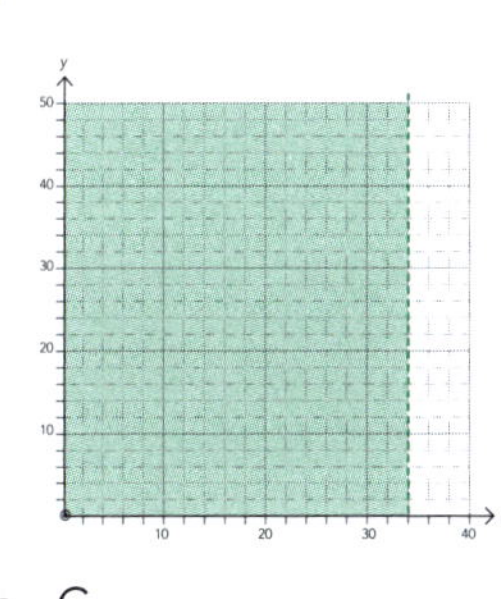

4 **a** D **b** C
c H **d** A
e G **f** B
g F **h** E

5 **a**

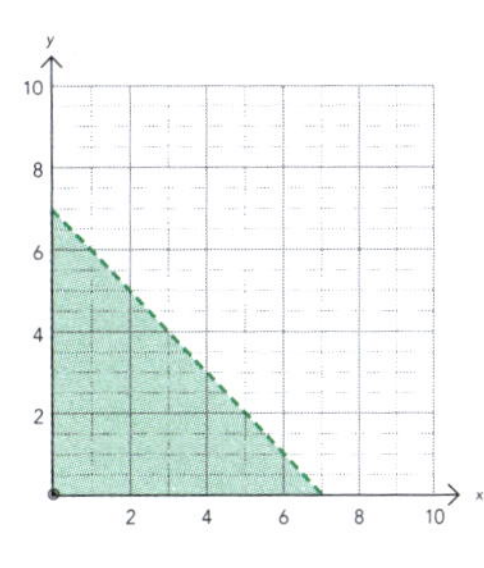

b

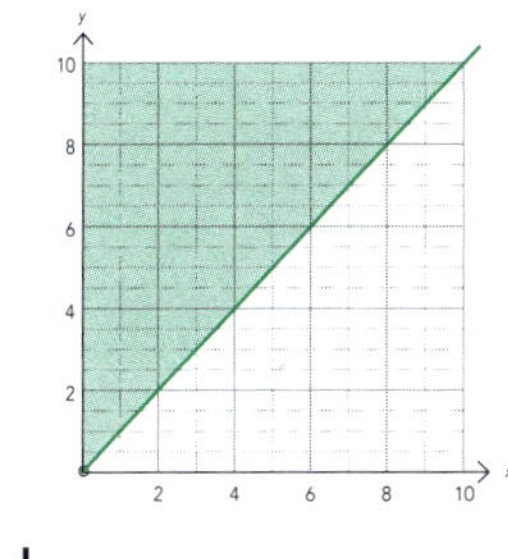

c

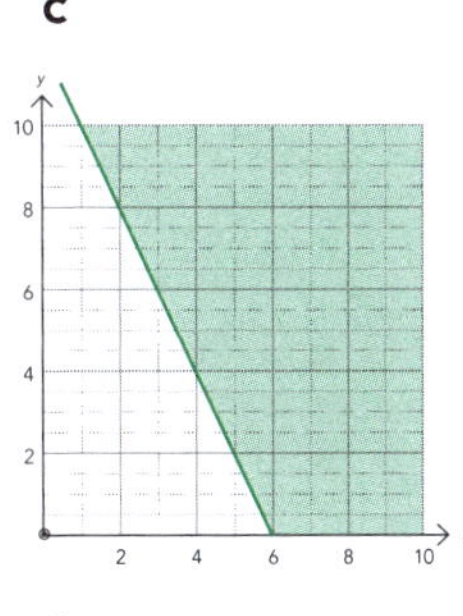

d

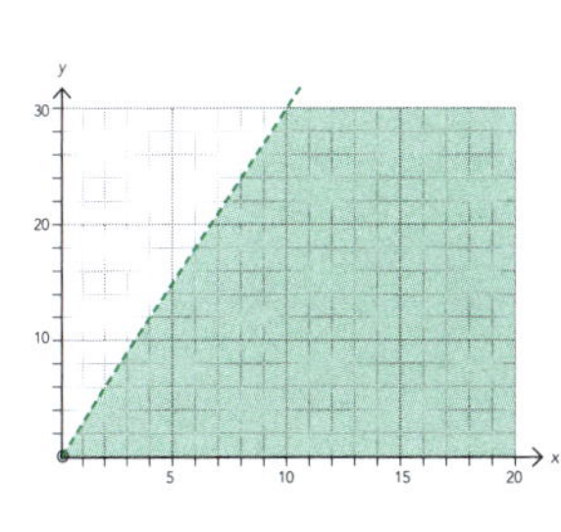

e

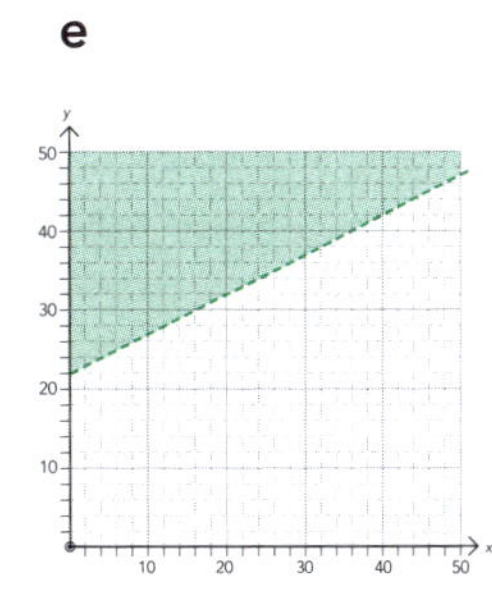

f

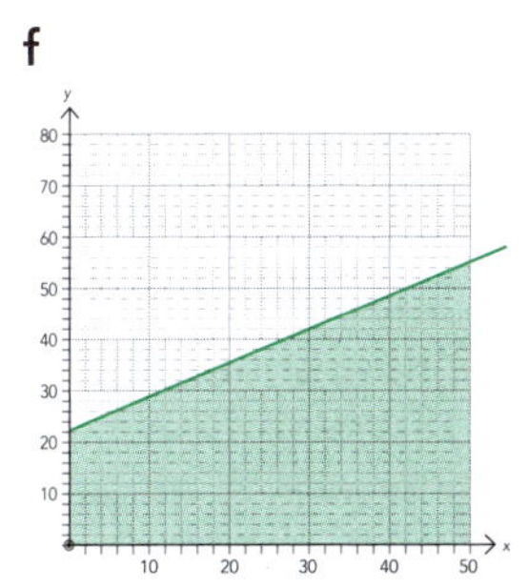

ISBN: 9780170389396

6

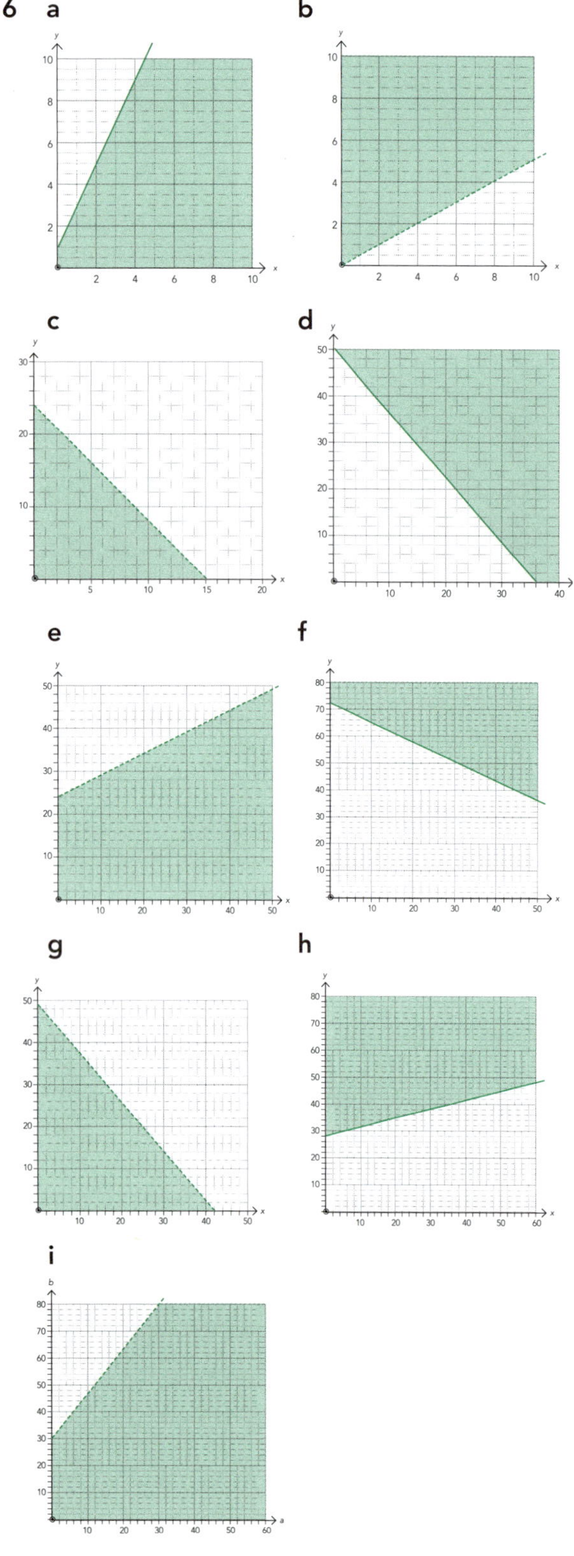

Putting it all together (pp. 46–55)

1 Vertices: (1, 9), (4, 9), (6, 6) and (1, 1)

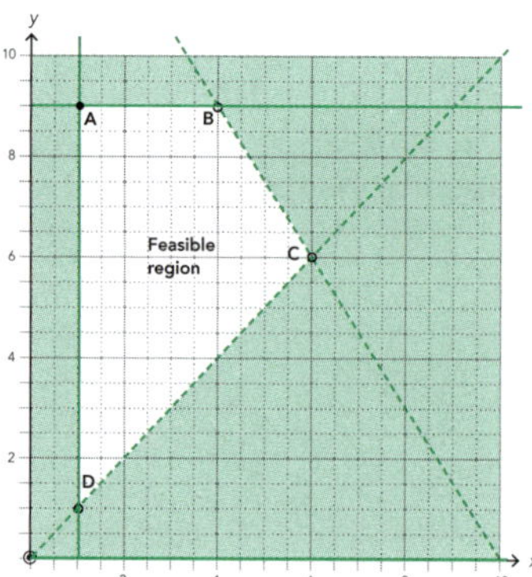

2 Vertices: (0, 10), (2, 6) and (6, 4)

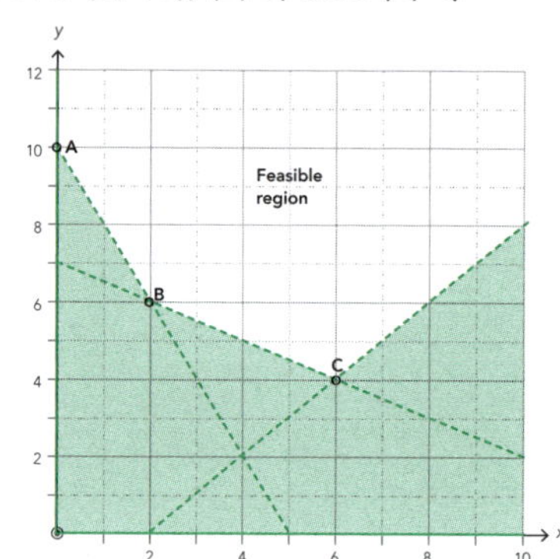

3 Vertices: $(\frac{2}{3}, 36)$, (30, 36), $(30, \frac{5}{2})$, (16, 6) and (8, 14)

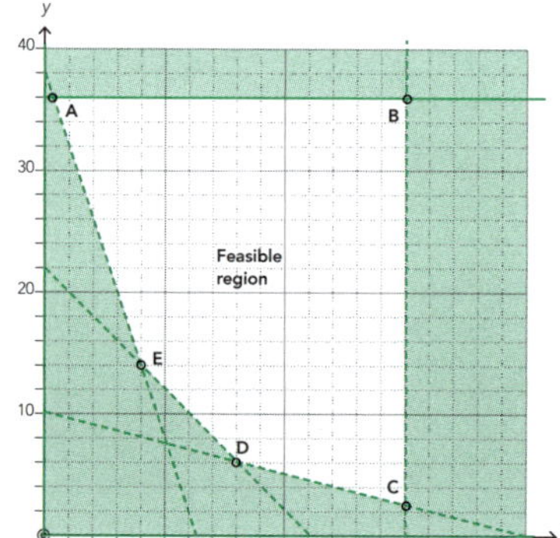

4 Vertices: (8,21.3), (24, 24), (36, 16), (40, 8) and (8, 8)

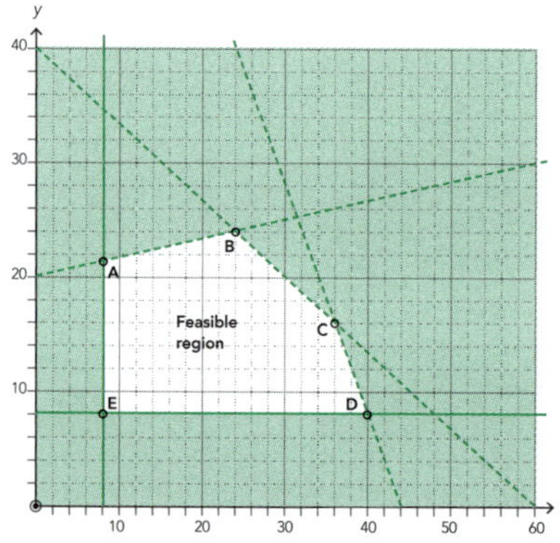

ISBN: 9780170389396

5 Vertices: (0, 96), (10, 56), (30, 32), (90, 2) and (100, 0)

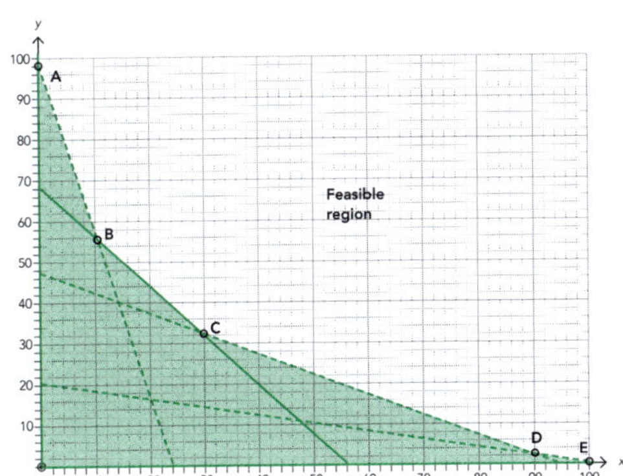

6 Vertices: (4, 92), (98, 92), (98, 0), (90, 0), (78, 6) and (16, 68)

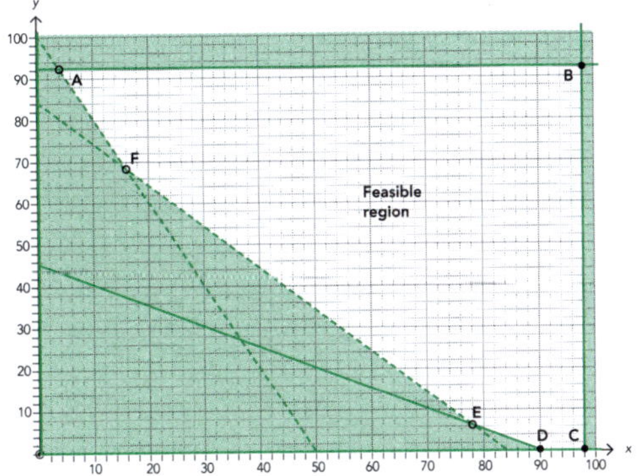

Writing inequations (pp. 56–59)

1 **a** $x \leq 15$ **b** $y > 40$
c $x + y \leq 70$ **d** $4x + 2y \leq 150$
e $y \geq 3x$ **f** $y - 30 = x$
g $y - x > 20$

2 **a** $y > x$ **b** $x \geq 150$
c $500 < y < 600$
d $50\,000 \leq 2400x + 800y \leq 60\,000$
e $x + y \leq 700$ **f** $y > 2x$
g $y \geq 3x$

3 $y < 3x$
$x + y \geq 20$
$x + y < 30$

4 $x + y \leq 1600$
$y \geq 5x$
$y - x \geq 900$

5 $x > 2y$
$75x + 85y \leq 1500$
$75x + 85y > 1000$
$75x \geq 85y + 400$

Linear programming (pp. 60–69)

1 Let x represent the number of first prizes and y represent the number of runner-up prizes.
Objective function: $P = 25x + 8y$

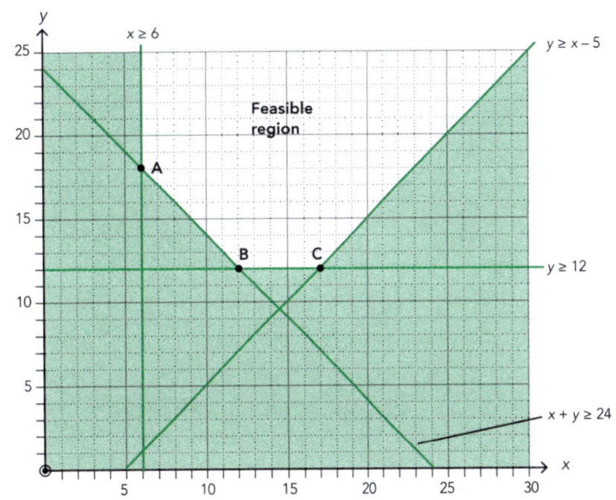

A: (6, 18): $P = 25(6) + 8(18) = 294$
B: (12, 12): $P = 25(12) + 8(12) = 396$
C: (17, 12): $P = 25(17) + 8(12) = 521$
∴ They should make 6 first-prize packs and 18 runner-up packs, which will cost the minimum amount of $294.

2 Let x represent the number of sunflower plants and y represent the number of poppy plants.
Objective function: $P = 2x + y$

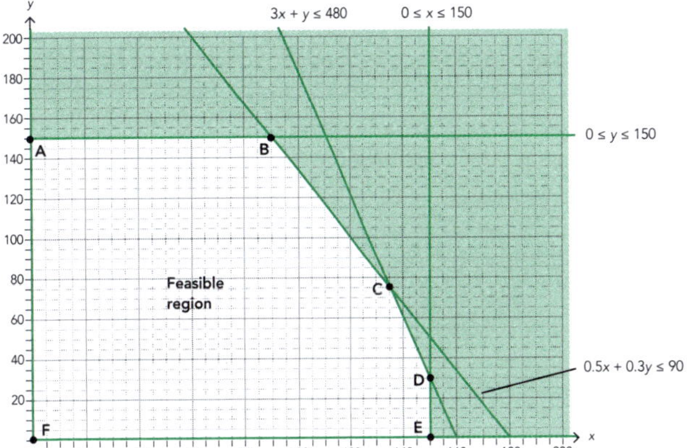

A: (0, 150) $P = 2(0) + (150) = 150$
B: (90, 150): $P = 2(90) + 1(150) = 330$
C: (135, 75): $P = 2(135) + 1(75) = 345$
D: (150, 30): $P = 2(150) + 1(30) = 330$
E: (150, 0) $P = 2(150) + 1(0) = 300$
F: (0, 0) $P = 2(0) + 1(0) = 0$
∴ The council should plant 135 sunflowers and 75 poppies in order to maximise the visual impact of the mass planting.

ISBN: 9780170389396

3 Let x represent the number of cars and y represent the number of vans.
Objective function: Number of students $= 4x + 10y$

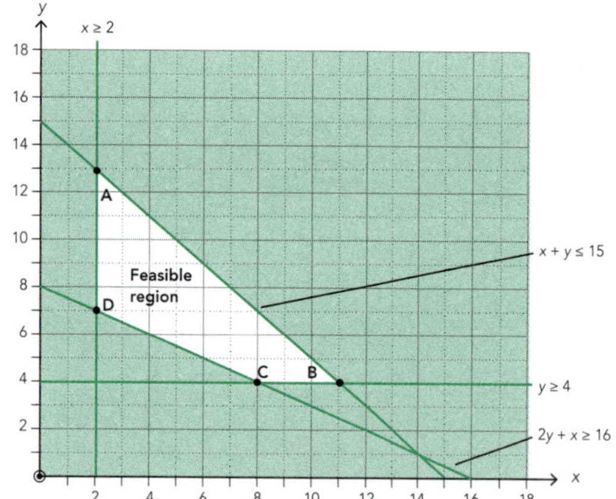

A: (2, 13): P = 4(2) + 10(13) = 138
B: (11, 4): P = 4(11) + 10(4) = 84
C: (8, 4): P = 4(8) + 10(4) = 72
D: (2, 7): P = 4(2) + 10(7) = 78

∴ If they take 8 cars and 4 vans, they can take a minimum of 72 students.
If they take 2 cars and 13 vans, then they can take a maximum of 138 students.
Limitation: in order to take 2 cars and 13 vans they would need to include 28 Year 13 students.

4 Let x represent the number of design A and y represent the number of design B.
Objective function: *Profit* $= 5x + 3y$

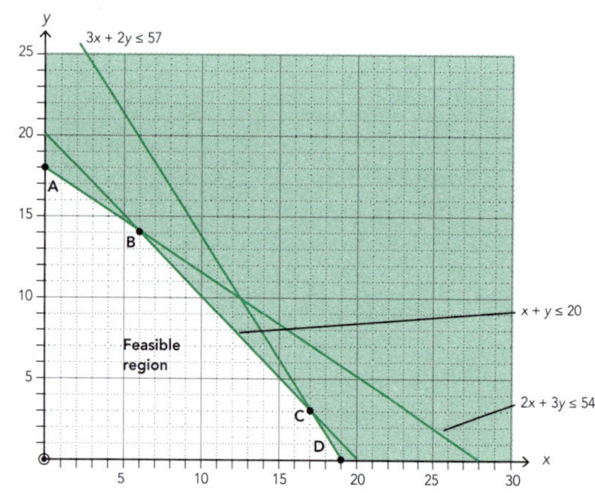

A: (0, 18): *P* = 5(0) + 3(18) = 54
B: (6, 14): *P* = 5(6) + 3(14) = 72
C: (17, 3): *P* = 5(17) + 3(3) = 94
D: (19, 0): *P* = 5(19) + 3(0) = 95

∴ She should make 19 of design A and none of design B, and this will make her a profit of $95. She will have 1 fern leaf and 16 sprigs of lavender left over.

How to deal with vertices that are not whole numbers (pp. 70-75)

1 Let x represent the number of hot-dogs and y represent the number of steak sandwiches.
Objective function: *Profit* $= 3x + 5y$

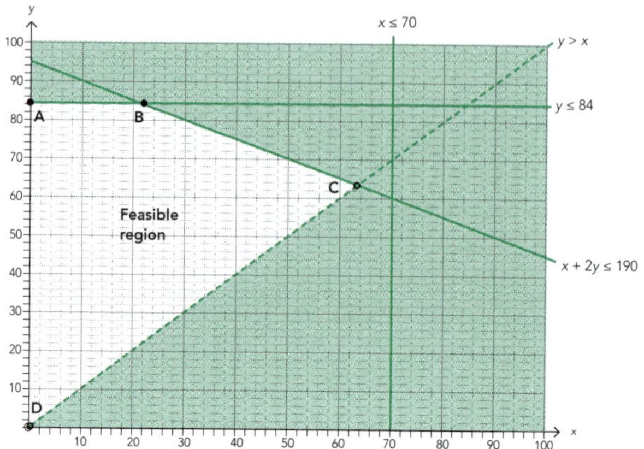

A: (0, 84): *P* = 3(0) + 5(84) = **420**
B: (22, 84): *P* = 3(22) + 5(84) = **486**
C: (63.3, 63.3): Must have integral points for number of hot-dogs and steak sandwiches.
∴ Consider points nearest to this, but inside the feasible region.

Consider (63, 63):

$y > x$:	63 > 63	✗

Consider (62, 63):

$y > x$:	63 > 62	✓
$x + 2y \le 190$	62 + 2(63) = 188	✓

P = 3(62) + 5(63) = **501**

Consider (62, 64):

$y > x$:	64 > 62	✓
$x + 2y \le 190$	62 + 2(64) = 190	✓

P = 3(62) + 5(64) = **506**

Consider (63, 64):

$y > x$:	64 > 63	✓
$x + 2y \le 190$	63 + 2(64) = 191	✗

D: (0, 0): Obviously no profit.

∴ They should make 62 hot-dogs and 64 steak sandwiches, and this will make them a profit of $506.

ISBN: 9780170389396

2 Let x represent the number of small seats and y represent the number of large seats.
Objective function: *Profit* = $30x + 25y$

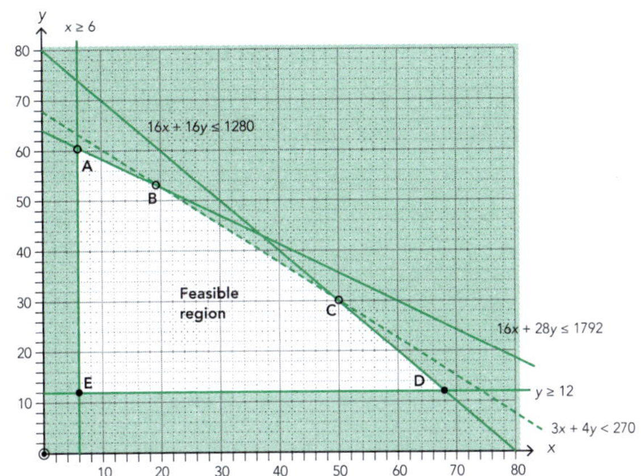

A: (6, 60.57): Must have integral points for number of small and large seats.
∴ Consider points nearest to this, but inside the feasible region.

Consider (6, 60):
$16x + 28y \leq 1792$: $16(6) + 28(60) = 1776$ ✓
$P = 30(6) + 25(60) =$ **1680**

B: (19.6, 52.8): Must have intergral points for number of small and large seats
∴ Consider points nearest to this, but inside the feasible region.

Consider (19, 52):
$3x + 4y < 270$ $\quad 3(19) + 4(52) = 265$ ✓
$16x + 28y \leq 1792$ $\quad 16(19) + 28(52) = 1760$ ✓

$P = 30(19) + 25(52) =$ **1870**

Consider (20, 52):
$3x + 4y < 270$ $\quad 3(20) + 4(52) = 268$ ✓
$16x + 28y \leq 1792$ $\quad 16(20) + 28(52) = 1776$ ✓

$P = 30(20) + 25(52) =$ **1900**

Consider (19, 53):
$3x + 4y < 270$ $\quad 3(19) + 4(53) = 269$ ✓
$16x + 28y \leq 1792$ $\quad 16(19) + 28(53) = 1788$ ✓

$P = 30(19) + 25(53) =$ **1895**

C: (50, 30): Not inside the feasible region because $3x + 4y$ **<** 270

Consider (49, 30):
$16x + 16y \leq 1280$ $\quad 16(49) + 16(30) = 1264$ ✓
$3x + 4y < 270$ $\quad 3(49) + 4(30) = 267$ ✓

$P = 30(49) + 25(30) =$ **2220**

Consider (50, 29):
$16x + 16y \leq 1280$ $\quad 16(50) + 16(29) = 1264$ ✓
$3x + 4y < 270$ $\quad 3(50) + 4(29) = 266$ ✓

$P = 30(50) + 25(29) =$ **2225**

D: (68,12): $P = 30(68) + 25(12) =$ **2340**
∴ They should make 68 small seats and 12 large seats, and this will make them a profit of \$2340.

Multiple solutions (pp. 76–83)

1 a Let x represent the number of mini-pizzas made and y represent the number of stuffed potatoes made.
Objective function: *Proft* = $0.5x + 0.75y$

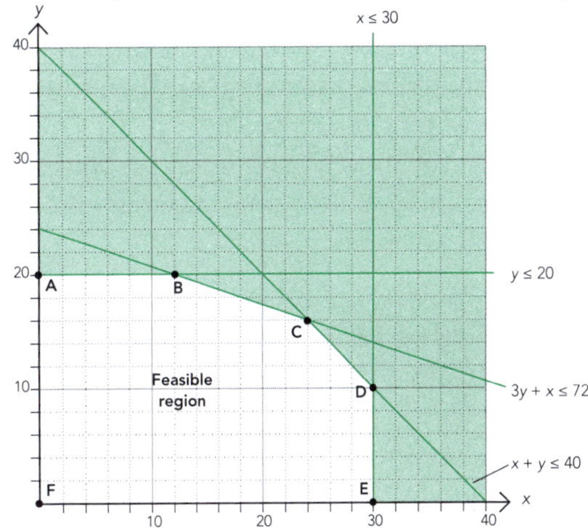

Value of the objective function at each vertex that could produce a maximum:

A: (0, 20): $P = 0.5(0) + 0.75(20) = \$15$
B: (12, 20): $P = 0.5(12) + 0.75(20) = \$21$
C: (24, 16): $P = 0.5(24) + 0.75(16) = \$24$
D: (30, 10): $P = 0.5(30) + 0.75(10) = \$22.50$
E: (30, 0): $P = 0.5(30) + 0.75(0) = \$15$
F: (0, 0): $P = 0.5(0) + 0.75(0) = \$0$

∴ The most profit of \$24 is made at point C (24, 16).

The maximum profit the canteen can make is \$24, which they get by making 24 mini-pizzas and 16 stuffed potatoes.

b New profits:
B: (12, 20): $P = 0.25(12) + 0.75(20) = \$18$
C: (24, 16): $P = 0.25(24) + 0.75(16) = \$18$
D: (30, 10): $P = 0.25(30) + 0.75(10) = \$15$
Because profit of \$18 is obtained at B and C, any integral point along the line connecting B and C ($x + 3y = 72$) will also give the same profit.
E.g. (18, 18).
∴ A maximum profit of \$18 occurs if they make (12 mini-pizzas, 20 stuffed potatoes), (15 mini-pizzas, 19 stuffed potatoes), (18 mini-pizzas, 18 stuffed potatoes), or (21 mini-pizzas, 17 stuffed potatoes), (24 mini-pizzas, 16 stuffed potatoes).

c The line $x + y = 40$ would have to be higher and it would need to pass through the point (30, 14).
The line needs to move up by 4 units
∴ $x + y = 44$.

Therefore they would have to increase their oven capacity to 44, an increase of 4 items.
The point (30, 14) would become a vertex of the new feasible region. Therefore the new profit would be
$P = 0.5(30) + 0.75(14) = \$25.50$. This is more than the original maximum profit of \$24, which occurred at C.

∴ The new maximum profit of \$25.50 would be from making 30 mini-pizzas and 14 stuffed potatoes.

d Objective function: $Profit = ax + 0.75y$
For the profit to be the same at points C and D, the gradient of the objective function would need to be the same as the gradient of the line that connects these points: $x + y = 40$.

$y = -x + 40$
∴ $m = -1$

$m = -1 \Rightarrow$ objective function is $Profit = 0.75x + 0.75y$, so they would need to make a 75-cent profit on both stuffed potatoes and mini-pizzas.
The new maximum profit becomes $0.75(24) + 0.75(16) = \$30$.
Other points that would generate this profit: any values between $x = 24$ and $x = 30$, provided the sum of x and y is 40.
Points: (24 mini-pizzas, 16 stuffed potatoes), (25 mini-pizzas, 15 stuffed potatoes), (26 mini-pizzas, 14 stuffed potatoes), …, (30 mini-pizzas, 10 stuffed potatoes).

2 a Let x represent the number of beginner's kits and y represent the number of intermediate kits.
Objective function: $P = 5x + 6y$

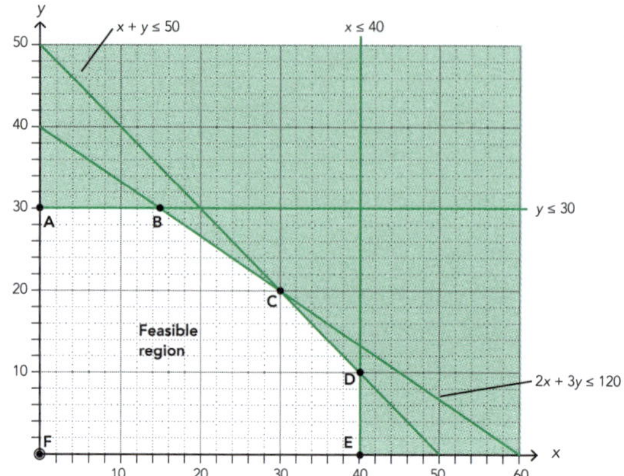

A: (0, 30)	$P = 5(0) + 6(30) = 180$
B: (15, 30):	$P = 5(15) + 6(30) = 255$
C: (30, 20):	$P = 5(30) + 6(20) = 270$
D: (40, 10):	$P = 5(40) + 6(10) = 260$
E: (40, 0)	$P = 5(40) + 6(0) = 200$
F: (0, 0)	$P = 5(0) + 6(0) = 0$

∴ He should assemble 30 beginner's kits and 20 intermediate kits, which will produce a maximum profit of \$270.

b New profits:

B: (15, 30):	$P = 6(15) + 6(30) = 270$
C: (30, 20):	$P = 6(30) + 6(20) = 300$
D: (40, 10):	$P = 6(40) + 6(10) = 300$

Because profit of \$300 is obtained at C and D, any integral point along the line connecting C and D ($x + y = 50$) will also give the same profit.

∴ A maximum profit of \$300 occurs if he makes (30 beginner's kits, 20 intermediate kits), (31 beginner's kits, 19 intermediate kits), (32 beginner's kits, 18 intermediate kits),…, (40 beginner's kits, 10 intermediate kits).

c The line $2x + 3y = 120$ would have to be higher and it would need to pass through the point (20, 30).
∴ The equation would need to be $2(20) + 3(30) = 130$, so $2x + 3y = 130$.
Therefore he would have to increase his time available by 10 minutes.

ISBN: 9780170389396

The point (20, 30) would become a vertex of the new feasible region. Therefore the new profit would be $P = 5(20) + 6(30) = \$280$. This is more than the original maximum profit of \$270, which occurred at C.

∴ The new maximum profit of \$280 would be from making 20 beginner's kits and 30 intermediate kits.

d Objective function:
Profit = 6x + ay
For the profit to be the same at points B and C, the gradient of the objective function would need to be the same as the gradient of the line that connects these points:

$$2x + 3y = 120$$
$$3y = -2x + 120$$
$$y = -\frac{2}{3}x + 40$$
$$\therefore m = -\frac{2}{3}$$

For the profit to be the same along the line BC, this gradient ($m = -\frac{3}{2}$) must apply to the profit equation also:

$$Profit = 6x + ay$$
$$ay = -6x + Profit$$
$$y = -\frac{6}{a} + \frac{Profit}{a}$$
$$m = -\frac{6}{a}$$
$$\therefore -\frac{6}{a} = -\frac{2}{3}$$
$$2a = 18$$
$$a = 9$$

∴ Eru would need to make a profit of \$9 on intermediate kits.

The new maximum profit becomes $6(15) + 9(30) = \$360$.
Other points that would generate this profit: any values between $x = 15$ and $x = 30$, provided $2x + 3y = 120$.
Points: (15 beginner's kits, 30 intermediate kits), (18 beginner's kits, 28 intermediate kits), (21 beginner's kits, 26 intermediate kits), (24 beginner's kits, 24 intermediate kits), (27 beginner's kits, 22 intermediate kits) or (30 beginner's kits, 20 intermediate kits) all prduce a profit of \$360.

Practice tasks (pp. 86–91)

Practice task one (pp. 86–87)

Let x represent the number of backstroke lengths and y represent the number of freestyle lengths.
Objective function: *Sponsorship* = 0.75x + 0.55y

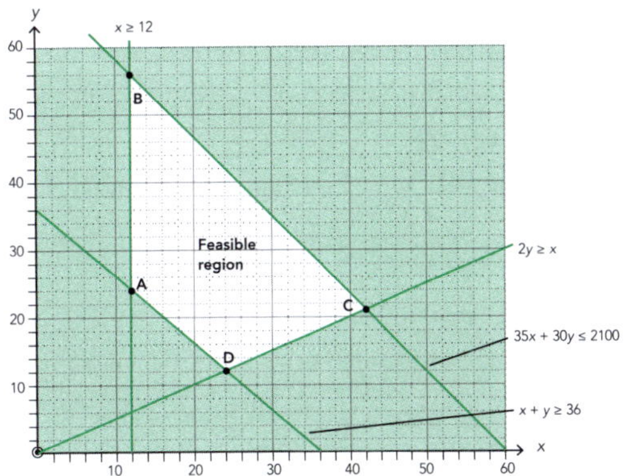

The only vertices that could produce a maximum sponsorship:

B: (12, 56): $S = 0.75(12) + 0.55(56) = 39.80$
C: (42, 21): $S = 0.75(42) + 0.55(21) = 43.05$
∴ She should do 42 lengths of backstroke and 21 lengths of freestyle.

Effect of changing coach's instructions:

Constraint equation becomes $3y \geq x$ rather than $2y \geq x$.

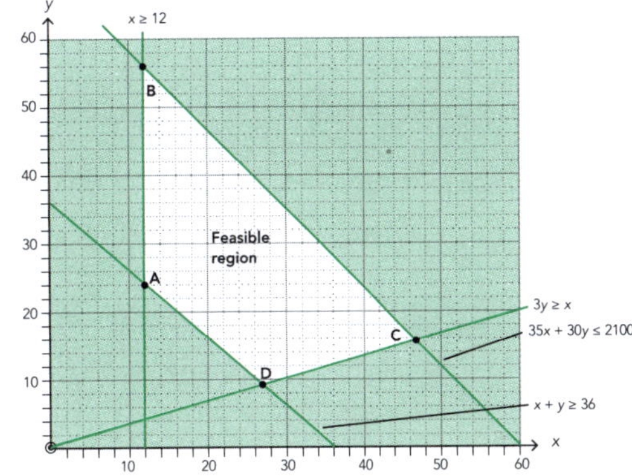

This changes the coordinates of point C (and D, but she could never earn maximum sponsorship at D because it involves swimming fewer of both types of length).

C: ($46.\dot{6}$, $15.\dot{5}$): Must have integral points for number of complete lengths swum ∴ consider points nearest to this, but inside the feasible region.

Consider (46, 15):

$35x + 30y \leq 2100$:	$35(46) + 30(15) = 2060$	✓
$3y \geq x$:	$3(15) = 45$	✗

Consider (47, 15):
$35x + 30y \le 2100$: $35(47) + 30(15) = 2095$ ✓
$3y \ge x$: $3(15) = 45$ ✗

Consider (46, 16):
$35x + 30y \le 2100$: $35(46) + 30(16) = 2090$ ✓
$3y \ge x$: $3(16) = 48$ ✓
$S = 0.75(46) + 0.55(16) =$ **43.30**

Consider (47, 16):
$35x + 30y \le 2100$: $35(47) + 30(16) = 2125$ ✗
$3y \ge x$: $3(16) = 48$ ✓

B: (19, 52):
Remains the same with sponsorship of $39.80.

∴ Changing the coach's requirement for doing at least 3 freestyle lengths for every backstroke length completed to 2 lengths of freestyle for every backstroke length means that she should do 46 backstroke lengths and 16 freestyle lengths. This will earn her $43.30.

Practice task two (pp. 88–89)
Let x represent the number of hens and y represent the number of ducks.
Objective function: $Income = 0.6x + y$

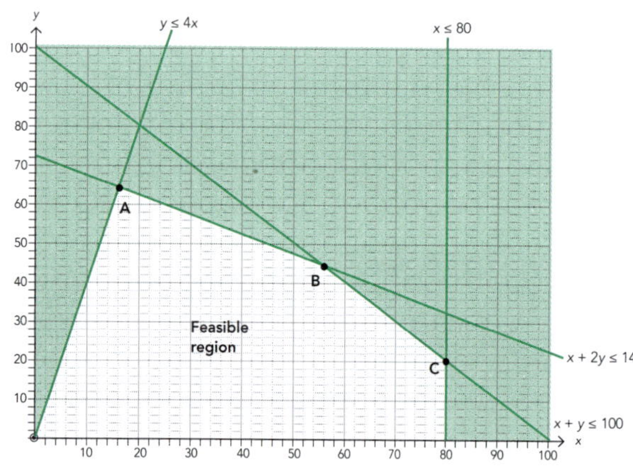

The only vertices that could produce a maximum income:
A: (16, 64): $I = 0.6(16) + 64 = 73.60$
B: (56, 44): $I = 0.6(56) + 44 = 77.60$
C: (80, 20): $I = 0.6(80) + 20 = 68.00$
∴ He should keep 56 hens and 44 ducks to produce a maximum income of $77.60 per day.

Increasing the price of duck eggs to $1.20
The only vertices that could produce a maximum income:
A: (16, 64): $I = 0.6(16) + 1.2(64) = 86.40$
B: (56, 44): $I = 0.6(56) + 1.2(44) = 86.40$
C: (80, 20): $I = 0.6(80) + 1.2(20) = 72.00$
Points A (16 hens and 64 ducks) and B (56 hens and 44 ducks) produce the same maximum income ($86.40), so any integral point along the line $x + 2y = 144$ between (16, 64) and (56, 44) will also produce the same maximum income.

Other points: (20 hens, 62 ducks), (24 hens, 60 ducks), (28 hens, 58 ducks), …, (52 hens, 46 ducks).

Increasing the budget for chicks and ducklings
The line $y = -\frac{1}{2}x + 72$ would have to be higher and it would need to pass through the point (20, 80).

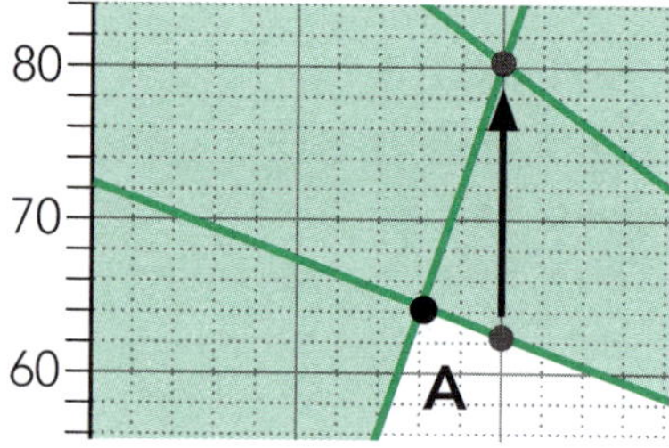

The line needs to move up by 18 units
∴ $y = -\frac{1}{2}x + 90$

$2y = -x + 180$
$x + 2y = 180$

Therefore he would have to increase his budget for chicks and ducklings to $180, an increase of $36.

The point (20, 80) would become a vertex of the new feasible region. Therefore if he kept 20 hens and 80 ducks, his maximum income would increase from $86.40 to 0.6(20) + 1.2(80) = $108.

ISBN: 9780170389396

Practice task three (pp. 90–91)

Let x represent the number of bags of ingredient A and y represent the number of bags of ingredient B.

Objective function: $Cost = 11x + 18y$

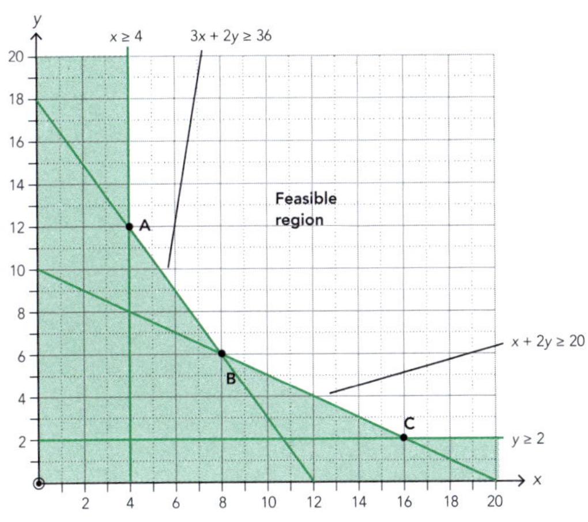

Vertices of the feasible region:

A: (4, 12): $C = 11(4) + 18(12) = 260$

B: (8, 6): $C = 11(8) + 18(6) = 196$

C: (16, 2): $C = 11(16) + 18(2) = 212$

∴ He should use 8 bags of ingredient A and 6 bags of ingredient B, and this will cost him a total of $196 per **batch**.

Minimum cost per bag of feed:

A: (4, 12): $C = 11(4) + 18(12) = 260 \Rightarrow$ cost per bag $= \frac{260}{4 + 12} = \$16.25$

B: (8, 6): $C = 11(8) + 18(6) = 196 \Rightarrow$ cost per bag $= \frac{196}{8 + 6} = \$14$

C: (16, 2): $C = 11(16) + 18(2) = 212 \Rightarrow$ cost per bag $= \frac{212}{16 + 2} = \$11.78$

∴ He should use 16 bags of ingredient A and 2 bags of ingredient B, and this will cost him a total of $11.78 per **bag**.

Increase protein content to more than 20 units:

A: (4, 12): $C = 11(4) + 18(12) =$ **260**

B: (8, 6): Not inside the feasible region because $x + 2y > 20$

Consider (8, 7):

$x + 2y > 20$: $8 + 2(7) = 22$ ✓

$3x + 2y \geq 36$ $3(8) + 2(7) = 38$ ✓

$C = 11(8) + 18(7) =$ **214**

Consider (9, 6):

$x + 2y > 20$: $9 + 2(6) = 21$ ✓

$3x + 2y \geq 36$ $3(9) + 2(6) = 39$ ✓

$C = 11(9) + 18(6) =$ **207**

C: (16, 2): Not inside the feasible region because $x + 2y > 20$

Consider (15, 3):

$x + 2y > 20$: $15 + 2(3) = 21$ ✓

$y \geq 2$ ✓

$C = 11(15) + 18(3) =$ **219**

Consider (17, 2):

$x + 2y > 20$: $17 + 2(2) = 21$ ✓

$y \geq 2$ ✓

$C = 11(17) + 18(2) =$ **223**

Minimum price per **batch** of feed:

He should use 9 bags of ingredient A and 6 bags of ingredient B, and this will cost him $207 per batch.

Minimum price per **bag** of feed:

B: (8, 7):

$C = 11(8) + 18(7) = 214$

$\Rightarrow$ cost per bag $= \frac{211}{8 + 7} = \$14.27$

B: (9, 6):

$C = 11(9) + 18(6) = 207$

$\Rightarrow$ cost per bag $= \frac{207}{9 + 6} = \$13.80$

C: (15, 3):

$C = 11(15) + 18(3) = 219$

$\Rightarrow$ cost per bag $= \frac{219}{15 + 3} = \$12.17$

C: (17, 2):

$C = 11(17) + 18(2) = 223$

$\Rightarrow$ cost per bag $= \frac{223}{17 + 2} = \$11.74$

He should use 17 bags of ingredient A and 2 bags of ingredient B, and this will cost him $11.74 per bag.